Exploring Perceptions from the Unknown 101:

A Speculative Scientific Journey Through Consciousness and the Hidden Universe

An Interdisciplinary Exploration Based on My Personal Journey

By Pearz Reagan

Abstract:

This narrative weaves together personal experiences, scientific exploration, and philosophical inquiry to explore the nature of reality, consciousness, and the universe. Through the lens of bipolar disorder and manic episodes, the author delves into profound questions about the interconnectedness of all things, the potential of black holes to transmit information across time and space, and the intersection of science and spirituality. This work invites readers to challenge their assumptions, broaden their understanding, and consider the possibilities that lie beyond the limits of current knowledge.

Copyright Page

© 2024 Pearz Reagan. All rights reserved.

No part of this publication may be reproduced, distributed, or transmitted in any form or by any means, including photocopying, recording, or other electronic or mechanical methods, without the prior written permission of the author, except in the case of brief quotations embodied in critical reviews and certain other non-commercial uses permitted by copyright law.

This work is a product of the author's personal exploration of speculative science, mental health, and spirituality. The views expressed in this work are the author's own and do not reflect those of any organizations with which the author may be affiliated.

Self-Published

First Edition: August, 2024

For permission requests, please contact the author at pearz.reagan@engineer.com.

Table of Contents

Exploring Perceptions from the Unknown 101:....................1
    Abstract:...1
        Copyright Page...2
Introduction...9
Chapter 1: The Journey Through Bipolar Disorder................15
Chapter 2: The Vision of "We Are All One".......................23
    Interpreting the "101" Message............................24
    Exploring the "101" Message................................29
    The Limitations of Human Understanding......................33
    The Potential for Unity and Healing.............................36
Chapter 3: Transmission of Information Through Black Holes...41
    Exploring Potential Mechanisms................................45
    The Ethical Implications of Information Transmission.........48
    Technological and Scientific Advancements....................52
Chapter 4: The DMT Molecule and Near-Death Experiences....56
    The Evolutionary Role of DMT.....................................60
Chapter 5: The Double-Slit Experiment............................66
Chapter 6: The Intersection of Science and Spirituality...........78
    Science and Spirituality in Practice: Bridging the Gap.........83
    The Future of Science and Spirituality: A Unified Vision.....88
Chapter 7: Black Holes, Quantum Mechanics, and the Reality...93
    The Role of Black Holes in the Evolution of the Universe....97
    The Limits of Human Knowledge and Black Holes............101
Chapter 8: Reflections on Mental Illness...........................107
    The Intersection of Mental Illness and the Spiritual..........112
    The Role of Science in Understanding Mental Illness.........117
Chapter 9: Consciousness, Reality, and the Universe.............123
Chapter 10: The Nature of Life, Death, and the Afterlife.........130
Chapter 11: Synthesis and Reflection on the Nature of Reality..137
Conclusion:..144
    Embracing the Mystery and Continuing the Journey..........144
APPENDICES..148
    Appendix A: Glossary of Key Terms.............................149

Appendix B: Further Reading and Resources..................154
Appendix C: Overview of Scientific Theories.................156
Appendix D: Explanation of Key Experiments................162
Appendix E: References...167
Acknowledgments..174
About the Author..175
Contact Information...176

Introduction

Throughout history, humanity has been driven by a profound desire to understand the universe, to unravel the mysteries that define our existence. This quest for knowledge has taken many forms—through religion, philosophy, and, more recently, the scientific method. While these approaches often seem to be at odds with one another, they share a common goal: to explain the nature of reality and our place within it.

For me, this journey has been deeply personal. My life has been shaped not only by a relentless curiosity but also by the challenges of living with bipolar disorder. This condition has brought with it periods of intense creativity and insight, as well as times of deep despair. During my manic episodes, my mind has often felt like it was operating on a different plane, making connections and experiencing realities that seem beyond the grasp of my everyday consciousness. These experiences have led me to explore some of the most profound questions about the universe, consciousness, and the interplay between science and spirituality.

This narrative is an attempt to document and share these explorations. It is not a conventional memoir, nor is it a purely academic treatise. Instead, it is a fusion of personal experiences, scientific inquiry, and philosophical reflection. My goal is not to provide definitive answers but to provoke thought, to challenge assumptions, and to open minds to the vast possibilities that lie beyond our current understanding.

The Challenge of Understanding the Universe

The universe is an incredibly complex and mysterious place. Even with all the advancements in science, we still only understand a fraction of the total knowledge within the seen and unseen universe. The observable universe, vast as it is, represents only a tiny portion of reality. There are forces and phenomena that we can barely detect, let alone fully comprehend. And yet, we persist in our quest to understand.

Science, with its rigorous methods and empirical approach, has brought us incredible insights. We've discovered the fundamental particles that make up matter, the forces that govern their interactions, and the vast scales of time and space that define the cosmos. But science also has its limitations. It is constrained by what we can observe, measure, and test. There are questions that science cannot yet answer, and perhaps never will.

Religion, on the other hand, offers a different approach to understanding. It provides a framework for interpreting the mysteries of existence, often through stories, symbols, and teachings passed down through generations. Religion seeks to explain the purpose and meaning behind the universe, offering comfort and guidance to those who seek it. But religious teachings are often based on ancient texts and traditions, which can make them seem out of step with modern scientific understanding.

My Personal Journey

My own journey into these questions has been deeply influenced by my experiences with bipolar disorder. My first documented major manic episode occurred on my 22nd birthday, leading to a diagnosis that would forever change my life. During these episodes, I often feel as though my mind is operating on a different level, accessing insights and making connections that seem to transcend ordinary reality. These experiences have led me to question the nature of consciousness, the possibility of accessing deeper truths about the universe, and the role that mental illness might play in this process.

This narrative is an attempt to reconcile these experiences with what I know from science and philosophy. It is a journey through the fabric of reality, exploring concepts like quantum entanglement, the nature of black holes, and the possibility of higher dimensions. It is also a reflection on the limitations of human understanding, the potential for convergence between science and spirituality, and the role of personal experience in shaping our view of the universe.

This narrative is an exploration of the boundaries between science, consciousness, and the mysteries of the universe. Inspired by personal experiences, particularly with bipolar disorder, it delves into speculative ideas that challenge conventional perspectives. Through this work, I aim to bridge the gap between science and spirituality, exploring how they may converge to offer new insights into the nature of reality.

However, the fear of stigma has often kept me from speaking freely about the things I have experienced. This stigma has added layers of social anxiety, which I likely felt long before my diagnosis. In various workplaces, where people became aware of my disorder, I was treated differently—often as if I were fragile or broken. This perception is something I struggle with deeply, as I do not want to be seen as less capable or competent due to my mental illness.

I have chosen to publish this work under a pseudonym because of my concern that future employers might refuse to hire me due to my disability. While I would love to be more open about my experiences, I believe that anonymity is the best way to share my thoughts at this time.

By sharing my journey and insights, I hope to challenge the stigma surrounding mental illness and encourage a greater understanding of the gifts and strengths that individuals with such conditions can offer. If people knew more about the mental health challenges faced by those around them, I believe they would foster deeper respect and understanding.

The Purpose of This Narrative

I do not claim to have all the answers, nor do I expect that everyone will agree with my interpretations. My hope is that by sharing my experiences and insights, I can encourage others to think more deeply about the profound questions that are presented. I invite readers to approach this narrative with both skepticism and an open mind, recognizing that the truth may lie somewhere beyond our current understanding.

Ultimately, this is a story about the human quest for knowledge and the mysteries that continue to elude us. It is about the limits of what we can know and the possibilities that lie beyond those limits. And it is about the personal journey of one individual trying to make sense of it all.

Disclaimer

This narrative is a unique blend of personal experiences, scientific exploration, and speculative thought. Due to the nature of my experiences, some of the ideas and concepts presented may venture into areas that could be considered pseudoscience, and at times, they may even resemble science fiction. My intent is not to present these ideas as established scientific fact, but rather to explore the possibilities and connections that I, personally, have found meaningful in my search for understanding.

I encourage readers to interpret the situations and concepts discussed in whatever way resonates most with them—whether as manifestations of mental illness, as a possible connection with the universe that may one day be explained by science, or as expressions of something more divine. I fully understand and respect any skepticism regarding these experiences and, in fact, advise readers to approach them with a critical eye. Skepticism is a valuable tool for navigating the complex and often uncharted territory that lies at the intersection of personal experience and scientific inquiry.

At the same time, I invite those who can bring themselves to be open-minded to consider these ideas as a different way of looking at the universe—a perspective that may offer new ways of thinking about existence, consciousness, and the mysteries that surround us.

Throughout this work, I rely heavily on scientific concepts in an attempt to explain my experiences to myself and to the reader. However, many of the connections I draw are speculative and would require further study and rigorous examination to find their place within the broader scientific community. My hope is that by sharing these thoughts, I might inspire curiosity and exploration, while also acknowledging the speculative nature of much of what is presented.

The Use of Artificial Intelligence in My Work

In my journey of exploring the intricate concepts of consciousness, science, and spirituality, I've been fortunate to engage with cutting-edge tools in artificial intelligence. My exploratory attitude toward AI led me to pursue a Master's of Science in Artificial Intelligence, where I am discovering numerous applications for AI in both my academic and personal pursuits. Among these, ChatGPT has been instrumental in helping me compile my thoughts and relate my experiences to broader scientific and philosophical perspectives. I even developed a specialized GPT to delve deeper into these topics, which is available below for others to use in their explorations. AI, for me, is not a shortcut but a powerful tool in our ongoing technological advancement.

Link to GPT:

https://chatgpt.com/g/g-Pa3Zsfd46-conceptualizing-the-unseen-universe

Chapter 1: The Journey Through Bipolar Disorder

Introduction to Bipolar Disorder

Bipolar disorder is a complex mental health condition characterized by episodes of mania and depression. These mood swings can range from extreme highs (mania) to deep lows (depression), affecting an individual's ability to function in daily life (Goodwin & Jamison, 2007). The disorder affects approximately 1-2% of the global population, though the exact prevalence can vary by region and diagnostic criteria (Merikangas et al., 2011).

The societal stigma surrounding mental illness, particularly bipolar disorder, is another significant challenge. Stigma can lead to social isolation, discrimination, and a reluctance to seek treatment, further exacerbating the difficulties faced by those with mental illness (Corrigan & Watson, 2002). Throughout my life, I have encountered situations where the stigma of bipolar disorder has led others to view me as fragile or broken, rather than as a competent and capable individual.

Living with bipolar disorder has been one of the most defining aspects of my life. It is a condition marked by extreme mood swings, from the highest highs of mania to the deepest lows of depression. For those who experience it, bipolar disorder is more than just a series of mood fluctuations—it is a journey through altered states of consciousness, where the boundaries of reality and imagination blur, and where profound insights can emerge alongside intense confusion.

My journey with bipolar disorder became known to me on my 22nd birthday, when I experienced my first major manic episode, although it is likely I was dealing with bipolar disorder years before. At the time, I was in my senior year of college, studying Biological Engineering. The stress of maintaining my grades which had slipped due to mental illness, navigating the complex social dynamics of college life, and facing the uncertainty of my future career had been building for months. For five consecutive nights, I averaged no more than one hour of sleep per night. By the time my birthday arrived, I had lost touch with reality.

The episode was severe enough that I was hospitalized, and it was during this hospitalization that I was diagnosed with bipolar disorder. This diagnosis explained many of the "energetic and creative" spells I had experienced throughout high school and college, as well as the periods of deep depression that often left me limited in my ability to function. It was a revelation that would shape my understanding of myself and my place in the world.

The Manic Experience: A Double-Edged Sword

Manic episodes, while often dangerous and disorienting, can also bring with them a sense of heightened awareness and creativity. During these periods, my mind feels as though it is operating on overdrive, making connections and generating ideas at a rapid pace. It is as if my brain is accessing a level of consciousness that is usually hidden from view—a place where the ordinary rules of reality no longer apply, and where the boundaries between the self and the universe begin to blur.

It is during these manic episodes that I have experienced some of the most profound insights of my life. These moments of clarity often feel like they are being "given" to me, as if I am tapping into an all-knowing knowledge base held by the universe itself. At the time, these experiences feel incredibly real, and I have no doubt of their authenticity. However, once the mania subsides, I am left to question whether these insights were genuine or simply the product of my altered mental state.

One of the most striking aspects of mania is the sense of connection it brings. During these episodes, I often feel a deep sense of interconnectedness with the universe and with all living things. This feeling of unity is accompanied by a sense of purpose—that there is something I am meant to do or share with the world. It is a powerful and intoxicating experience, but it is also one that must be approached with caution.

The Burden of Depression

On the other side of the bipolar spectrum lies depression—a state of mind that is as debilitating as mania is exhilarating. During depressive episodes, the energy and creativity that characterize mania are replaced by a deep sense of despair and hopelessness. It is a time when the world seems gray and lifeless, and when even the simplest tasks can feel insurmountable.

In these moments, I often find myself reflecting on my manic experiences, trying to make sense of the insights I received and questioning whether they were real or simply the result of my mental illness. Depression brings with it a profound sense of self-loathing and inadequacy, making it difficult to see any value in the experiences I had during mania.

Despite the challenges of living with bipolar disorder, I have learned to find solace in the idea that there may be a purpose to my condition. While I despise the disorder and the toll it takes on my life, I also recognize that it may have given me access to insights and experiences that I am meant to share with the world. Perhaps there is an underlying meaning to my life beyond simply managing my condition—perhaps I am supposed to help others see the world in a new way.

Coping with Bipolar Disorder

Over time, I have developed strategies for coping with bipolar disorder. I have learned to recognize the early signs of mania and depression, and I have developed techniques for managing these episodes before they spiral out of control. Medication, therapy, and a strong support network have all been crucial in helping me navigate the challenges of living with this condition. However, my body seems to remain resistant to accessing the full benefits of the medications I've been prescribed so far, so I'm still in a decade long process of trial and error in finding the medication regiment right for me.

One of the most important lessons I have learned is the value of self-awareness. By understanding my triggers and learning to recognize the patterns in my mood swings, I have been able to take proactive steps to maintain my mental health. This has allowed me to continue pursuing my passions and interests, even in the face of the challenges posed by bipolar disorder. Yet, but learning self-awareness has also benefited other parts of my life as well allowing me the ability to better myself in many areas of my life as time goes on.

Another key aspect of coping with bipolar disorder is learning to accept and embrace the condition as part of who I am. While it can be tempting to see the disorder as something to be fought against, I have come to realize that it is also a source of strength and insight. By accepting my condition and learning to work with it, rather than against it, I have been able to channel the energy and creativity of my manic episodes into something positive.

The Intersection of Mental Illness and Insight

One of the most challenging aspects of living with bipolar disorder is the difficulty in distinguishing between genuine insights and the distortions caused by my mental illness. During my manic episodes, I have experienced what I can only describe as moments of profound clarity—times when I felt as though the universe was revealing its secrets to me, offering glimpses of a reality far beyond what we normally perceive. These experiences have been accompanied by a strong sense of certainty, a feeling that what I was seeing and understanding was true, even if it defied logic or reason.

However, the very nature of bipolar disorder makes it impossible to be entirely certain of the authenticity of these experiences. While in the throes of mania, everything feels real, but once the episode passes, doubt begins to creep in. Was I truly accessing some deeper level of understanding, or were these experiences simply the result of my mind being in overdrive, creating elaborate delusions out of thin air? This ambiguity is one of the most difficult aspects of my condition to reconcile.

Despite this uncertainty, I have come to believe that there may be something more to these experiences. Even if they are influenced by my mental illness, that does not necessarily mean they are without value. After all, many of the greatest breakthroughs in science and philosophy have come from individuals who thought outside the conventional bounds of reality—who questioned the status quo and dared to explore the unknown. Perhaps my experiences are simply another way of accessing that same creative potential, albeit through a different and more challenging path.

A Premonition of Termination

One particular experience stands out as both a confirmation of this belief and a source of ongoing uncertainty. During a manic episode, I experienced what I can only describe as a premonition about my future. It was not a vision in the traditional sense, but rather a series of ideas and concepts that were conveyed to me, almost as if they were being transmitted directly into my mind by an external source.

At the time, I was working for a company I will refer to as "Mayan Millers." I had always considered myself a model employee, dedicated to my work and committed to the success of the company. However, during this episode, I received a vivid "message" that I would no longer be working at Mayan Millers in a few months' time. This seemed impossible to me, yet the message was clear: I would have a confrontation with a superior, resulting in my termination, after which I would find myself working for another company, which I will call "US Quality Foods."

The most unsettling part of this premonition was the level of detail it provided. It predicted that I would lose my temper while within a manic state and "cuss out" the superintendent of the manufacturing plant. This would lead to my dismissal and a subsequent lawsuit for workplace discrimination. Following this, I would be unable to communicate with anyone from Mayan Millers due to legal constraints, and I would soon find myself employed at US Quality Foods.

Several months later, to my astonishment, these events unfolded exactly as predicted. The confrontation, the termination, the lawsuit, and my eventual employment at US Quality Foods all occurred as described. This experience has left me questioning the nature of reality, the boundaries of human perception, and the potential for our minds to access information that seems beyond the scope of ordinary understanding.

My wife suggested that it is possible my memory of the situation has been altered over time, that perhaps I have retroactively imposed this narrative onto the events that transpired. While this is a reasonable explanation, I am known for having an excellent memory, and the accuracy of the premonition is difficult for me to dismiss as mere coincidence.

Reconciling Science and Personal Experience

The experience of this premonition has brought me face to face with some of the most profound questions about the nature of reality and the limits of human understanding. On one hand, it is entirely possible that this was simply a case of my mind playing tricks on me, that the premonition was nothing more than a delusion caused by my mental illness. On the other hand, the specificity and accuracy of the prediction make it difficult to dismiss the possibility that there was something more at work—something that defies our current understanding of how the mind and the universe interact.

This experience has led me to explore various scientific and philosophical concepts in an attempt to find some kind of explanation. I have delved into theories of quantum mechanics, particularly the ideas surrounding quantum entanglement and the possibility of information being transmitted across time and space.

At the same time, I am acutely aware of the need to approach these ideas with caution. The boundary between genuine insight and delusion is a thin one, and it is important to remain grounded in reality while exploring the possibilities that lie beyond it. I do not claim to have all the answers, and I fully acknowledge the role that my mental illness may play in shaping my perceptions. However, I believe that these experiences, whether real or imagined, have value in the broader quest to understand the mysteries of consciousness and the universe.

Chapter 2: The Vision of "We Are All One"

Introduction to the Vision

One of the most profound and transformative experiences of my life occurred during a manic episode, when I received a vision that conveyed a message of universal interconnectedness. The message was simple yet powerful: "We are all One." This idea, symbolized by the number "101," has deeply influenced my understanding of humanity, society, and the universe as a whole.

The number 101 has also been seen in various cultures as a symbol of unity and balance (Ellis, 2004). In binary code, the sequence "101" represents the number 5, which is often associated with balance and change, emphasizing the duality and interconnectedness of all things (Knuth, 1997).

The vision did not come to me in the form of spoken words or images but as a deep, intuitive understanding that seemed to be "downloaded" into my consciousness. It was as if I had tapped into a universal truth, one that transcended the limitations of language and thought. The number "101" was repeatedly emphasized, with the first "1" representing unity, the "0" signifying perpetuity or infinity, and the final "1" reaffirming the concept of unity. Together, these symbols conveyed the idea that all beings and things in the universe are fundamentally connected, originating from a single source and eventually returning to it.

The concept of a journey from unity to unity is common in many philosophical traditions, where it represents the cycle of life and the return to the original source. This interpretation suggests that the "101" message could symbolize the completion of a cycle that brings together beginning, middle, and end—birth, life, and death, or creation, preservation, and dissolution (Capra, 1975).

Interpreting the "101" Message

The "101" message, as I have come to understand it, represents the cyclical nature of existence. It suggests that all life, consciousness, and matter in the universe are interconnected and that this interconnectedness is both eternal and unbreakable. The first "1" signifies the beginning—unity, where everything originates from a single point or entity. The "0" in the middle represents the infinite journey, the process of existence where individual entities explore and interact with the universe. The final "1" signifies the return to unity, where everything eventually comes back together, completing the cycle (West, 2011; Ellis, 2004).

I've attempted to look at the meaning of this more as time went on. At the time, I understood it as binary, but I was unsure why I was not receiving more digits to complete a more complex message since there was more complicated conceptual content conveyed with this message. However, after researching the possible significance of the number "101," I found some interesting findings:

Numerical Interpretation:

- **101 as the Number 5:** In numerology, 1 + 0 + 1 equals 2, symbolizing balance and harmony, which reflects the cyclical nature of existence where dualities merge into a unified whole. Alternatively, in binary, "101" represents the number 5, which is often associated with change, adaptability, and dynamic balance in numerology, suggesting that the universe constantly evolves but maintains a balance (West, 2011).

Binary Code:

- **Binary Implications:** "101" is a binary number, a foundational concept in computing and digital logic. In binary code, "101" represents the number 5, emphasizing the balance between on and off, presence and absence, or being and non-being. This duality can be seen as a metaphor for the interconnectedness of all things, where opposites coexist and contribute to the overall balance of the universe (Knuth, 1997).

- At the time I received this message, I understood that it was binary code. I was perplexed as to the reason I wasn't receiving more digits, as a longer series is often required in order to create a more complex message. However, the three digits in themselves revealed quite a bit of significance in a simplistic manner.

Significance of Three:

- **Trinity and Triplicity:** The sequence "101" consists of three digits, and the number three often symbolizes completeness, stability, and the synthesis of opposites in many philosophical and spiritual traditions. This interpretation suggests that "101" could represent the completion of a cycle that brings together beginning, middle, and end—birth, life, and death, or creation, preservation, and dissolution (Ellis, 2004; Capra, 1975).

Foundation of Knowledge:

- **Academic Subfix:** The term "101" is commonly used in academia to designate introductory courses that lay the foundation for further study in a particular field. This could imply that the "101" message is foundational knowledge—a basic yet profound truth that underlies more complex understandings of the universe and our place within it (West, 2011).

Symbolic Journey:

- **Journey from Unity to Unity:** The sequence might also represent a journey—from unity, through a period of understanding or enlightenment (the "0"), back to unity. It can symbolize the cyclical nature of existence, where everything begins and ends with the same unified source (Capra, 1975).

Palindrome Significance:

- **Symmetry and Reflection:** The fact that "101" is a palindrome, reading the same forward and backward, symbolizes symmetry, balance, and the concept of duality or reflection. This could suggest that the message of "101" reflects the idea that the beginning and the end are mirror images of each other, reinforcing the cyclical nature of existence. The journey through life (represented by "0") leads back to the original unity (the final "1"), just as one might return to the beginning after completing a cycle (West, 2011; Ellis, 2004).

Angel Number:

- **101 as an Angel Number**: The number "101" is often referred to as an "Angel Number" in spiritual traditions, symbolizing the presence of divine guidance and the importance of maintaining a positive outlook during transitions and new beginnings. This interpretation aligns with the idea that "101" represents a journey from unity through understanding and back to unity, guided by higher spiritual forces (Doreen Virtue, 2005).

The Implications for Society

The "101" message has made me acutely aware of how human beings often get caught up in the superficial distinctions that separate us, rather than focusing on the deeper connections that unite us. In many ways, our society is structured to emphasize these differences, particularly in the realms of politics and religion.

> **Politics and Division:** Politics, as it is practiced today, often seems more focused on dividing people than on bringing them together. The bipartisan system, particularly in countries like the United States, tends to create an "us versus them" mentality, where individuals are forced to align themselves with one of two major political parties. This binary approach oversimplifies the complexities of human

beliefs and values, reducing them to a one-dimensional spectrum. However, political beliefs actually exist on a multidimensional plane, with a vast array of perspectives that cannot be neatly categorized into just two opposing sides. By focusing on these superficial distinctions, we lose sight of the common goals and values that should unite us as a society.

Religion and Segmentation: Religion, too, often plays a role in segmenting people into distinct groups based on shared beliefs and practices. While many religions preach messages of love, peace, and unity, they can also create boundaries that separate individuals from those who do not share the same faith. These divisions can lead to misunderstandings, conflicts, and a lack of empathy for those outside one's own religious community. The "101" message challenges this segmentation, suggesting that all spiritual paths are ultimately leading toward the same truth: that we are all part of a single, interconnected whole.

A Manic Episode and Lessons from the Universe

During this particularly intense manic episode, I felt as though my mind was being accessed by an external intelligence, receiving lessons about the universe that were beyond my ordinary understanding. These lessons seemed to branch out from a single entity, an all-encompassing source from which everything in the universe originates. The experience was both exhilarating and overwhelming, as if I were being shown the very fabric of reality and my place within it.

In this state, the message of "We are all One" was reinforced. I was shown that all things in the universe are connected, that every action has a ripple effect that influences the whole. It was as if the universe was a vast, intricate web, and each of us is a thread within that web. The more I understood this interconnectedness, the more I realized the importance of compassion, empathy, and understanding in our interactions with others.

This episode left me with a sense of awe and reverence for the complexity and beauty of the universe. It also deepened my belief that the divisions we create among ourselves are artificial and ultimately harmful. If we could all see the world through the lens of "101," recognizing our fundamental unity, I believe that many of the conflicts and misunderstandings that plague our society could be resolved.

Exploring the "101" Message through Science and Philosophy

The "101" message is not just a spiritual or philosophical concept; it also resonates with certain scientific theories that explore the nature of reality and interconnectedness. By delving deeper into these scientific concepts, we can gain a better understanding of how the vision of "We are all One" might align with our current understanding of the universe.

Quantum Entanglement: A Scientific Parallel

One of the most intriguing parallels to the "101" message in modern science is the phenomenon of quantum entanglement. Quantum entanglement is a concept in quantum mechanics where two or more particles become linked in such a way that the state of one particle is directly connected to the state of the other, regardless of the distance between them. This connection is instantaneous, meaning that changes to one particle are reflected in the other, even if they are light-years apart. This phenomenon challenges our classical understanding of physics and suggests that at a fundamental level, everything in the universe may be interconnected.

In some ways, quantum entanglement mirrors the idea that "We are all One." Just as entangled particles are inextricably linked, the vision suggests that all beings and things in the universe are connected, sharing a common origin and ultimately converging back into unity. The idea that particles can influence each other across vast distances aligns with the notion that our actions and thoughts have far-reaching consequences, affecting not only ourselves but also the entire web of existence.

While quantum entanglement is still not fully understood, it represents a profound shift in our understanding of the universe. It suggests that the fabric of reality is far more interconnected and complex than we previously imagined, and it opens up new possibilities for exploring the nature of consciousness and the universe.

The Flatland Thought Experiment: Perceiving Higher Dimensions

To further explore the limitations of our understanding and perception, consider the Flatland thought experiment. Originally conceived by Edwin A. Abbott in his novella *Flatland: A Romance of Many Dimensions*, the experiment imagines a two-dimensional world inhabited by flat beings who can only perceive length and width, but not height (Abbott, 1884). These beings live their lives in a flat plane, unable to comprehend the existence of a third dimension.

Now, imagine that a three-dimensional being—a sphere, for example—interacts with the inhabitants of Flatland. To the Flatlanders, the sphere would appear as a series of two-dimensional shapes, circles in this example, as it passes through their plane, but they would be unable to perceive its full form. The sphere could see everything in Flatland, understanding its entirety, but the Flatlanders could only perceive the sphere's interactions with their world as inexplicable phenomena.

This thought experiment serves as a powerful metaphor for our own limitations in perceiving higher dimensions. Just as the Flatlanders cannot comprehend the third dimension, we may be unable to perceive or understand dimensions beyond our own four-dimensional reality (three spatial dimensions plus time). The complexity of the universe, including phenomena like quantum entanglement and black holes, might be manifestations of these higher dimensions interacting with our own.

The "101" message could be interpreted as an insight into these higher dimensions—an understanding that goes beyond our conventional perception and suggests a more interconnected reality. It challenges us to think beyond the limitations of our senses and to consider the possibility that there are aspects of the universe that we are simply not equipped to perceive directly. Just as the Flatlanders cannot grasp the concept of height, we may struggle to comprehend the true nature of the universe.

Alternatively, within the context of the Flatland thought experiment, I could be seen as a flat shape that was granted the opportunity to interact with a three-dimensional element. While I may not have had the ability to fully access all the knowledge it possessed, I was able to perceive three-dimensional information in a two-dimensional format, allowing me to receive at least part of the message in a way that I could understand.

The Philosophical Implications of "101"

The "101" message also carries significant philosophical implications, particularly in the context of metaphysics and existentialism. It raises questions about the nature of existence, the interconnectedness of all things, and the purpose of life. If we are all connected, originating from the same source and eventually returning to it, what does that mean for our individual lives and the choices we make? (Sartre, 1943; Heidegger, 1927).

One possible interpretation is that the "101" message emphasizes the importance of living in harmony with others and the universe. If we are all part of the same whole, then harming others is, in a sense, harming ourselves. This perspective aligns with many ethical and spiritual teachings that advocate for compassion, empathy, and kindness as fundamental principles of human interaction (Bhagavad Gita; Singer, 1981).

Moreover, the idea that we are all interconnected suggests that our actions have far-reaching consequences, both in this life and beyond.

It challenges us to consider the impact of our decisions not only on our immediate surroundings but also on the broader web of existence. This understanding could inspire a sense of responsibility and purpose, encouraging us to live in a way that contributes positively to the world and the universe as a whole (Whitehead, 1929).

The Limitations of Human Understanding

The "101" message and the concepts we've explored thus far—quantum entanglement, the Flatland thought experiment, and the philosophical implications of interconnectedness—highlight a critical theme in this narrative: the limitations of human understanding. While these ideas provide a glimpse into the possible nature of reality, they also underscore the fact that there is much we do not know, and perhaps cannot know, about the universe (Kant, 1781).

The Incompleteness of Science

Science, as powerful and transformative as it is, has its limitations. Our current scientific understanding is based on observable phenomena, empirical evidence, and the application of mathematical models to describe the workings of the universe. However, the universe is vast and complex, and much of it remains beyond our observational capabilities. Dark matter, dark energy, the true nature of black holes—these are just a few of the mysteries that continue to elude us (Hawking, 1988).

The "101" message suggests that there is a deeper layer of reality that we have yet to access fully. It hints at the possibility that our current scientific models may be incomplete or that they describe only a small fraction of the true nature of existence. While quantum mechanics has opened the door to understanding the interconnectedness of particles and the strange behavior of matter at the smallest scales, it also raises more questions than it answers (Feynman, 1985).

This incompleteness is not a failure of science, but rather a reminder of its provisional nature. Science is a tool—a powerful one—but it is not the ultimate arbiter of truth. It is constantly evolving, driven by new discoveries and insights that challenge existing paradigms (Popper, 1934; Kuhn, 1962). The "101" message invites us to keep an open mind, to remain humble in the face of the unknown, and to recognize that there may be aspects of reality that are beyond the reach of our current methods of inquiry (Feyerabend, 1975).

The Role of Intuition and Experience

While science relies on empirical evidence, there are other ways of knowing that should not be dismissed. Intuition, personal experience, and the insights gained through altered states of consciousness—such as those experienced during manic episodes—can provide valuable perspectives on the nature of reality. These forms of knowledge are often subjective and difficult to quantify, but that does not necessarily diminish their importance (Bergson, 1934; Jung, 1963).

The "101" message, as I received it, was an intuitive understanding that felt as real and true as any scientific fact. It was a message that resonated deeply with my sense of self and my place in the universe, and it provided a framework for interpreting my experiences and the world around me. While I recognize that these insights may be influenced by my mental illness, I also believe that they offer a different kind of truth—one that complements, rather than contradicts, the knowledge gained through science (Wilber, 1998).

In many ways, the "101" message serves as a bridge between science and spirituality, between the empirical and the intuitive. It suggests that there is a place for both forms of knowledge in our quest to understand the universe, and that by integrating them, we may come closer to grasping the full scope of reality (Wilber, 1998).

A Call for Open-Mindedness

The message of "We are all One" is also a call for open-mindedness—a reminder that our understanding of the universe is always evolving and that we must be willing to consider new ideas and perspectives, even if they challenge our current beliefs. This is particularly important in a world where dogma—whether religious, scientific, or ideological—can easily take hold, closing off avenues of inquiry and stifling intellectual growth (James, 1897; Krishnamurti, 1969).

In science, this means being open to the possibility that our current theories may one day be replaced by more accurate models, and that phenomena we cannot yet explain may hold the key to new discoveries (Feyerabend, 1975; Sheldrake, 2012). It means recognizing the limitations of our tools and methods and acknowledging that there may be aspects of reality that we are not yet capable of understanding (Feyerabend, 1975).

In spirituality, it means being open to the insights and experiences of others, even if they do not align with our own beliefs. It means recognizing the value of personal experience, intuition, and the sense of connection that transcends individual existence. It also means being willing to question established doctrines and to explore new ways of thinking about the divine, the universe, and our place within it (James, 1897; Krishnamurti, 1969).

The "101" message challenges us to transcend the boundaries of conventional thought and to embrace a more holistic understanding of reality—one that integrates the empirical with the intuitive, the scientific with the spiritual, and the individual with the universal (Wilber, 1998).

The Potential for Unity and Healing

The "101" message, with its emphasis on universal interconnectedness, offers a powerful vision for unity and healing in both individual lives and society as a whole. In a world increasingly divided by political, religious, and cultural differences, the idea that "We are all One" serves as a reminder of our shared humanity and the deep connections that bind us together, regardless of our superficial differences (Hanh, 1998; Macy, 1991).

Healing Through Understanding

One of the most profound implications of the "101" message is its potential to foster healing—both on a personal level and within society at large. By recognizing our interconnectedness, we can begin to see others not as adversaries or strangers, but as extensions of ourselves, deserving of the same respect, compassion, and understanding that we seek for ourselves (Chödrön, 1997; Tutu, 1999).

On a personal level, this recognition can be deeply transformative. It can help us to overcome feelings of isolation and alienation, replacing them with a sense of belonging and purpose. Understanding that we are part of a larger whole can provide comfort and reassurance, especially during times of struggle or uncertainty. It can also encourage us to act with greater empathy and kindness, knowing that our actions have a ripple effect that extends far beyond our immediate surroundings (Chödrön, 1997).

In society, the "101" message has the potential to bridge divides and bring people together in pursuit of common goals. When we recognize that our well-being is inextricably linked to the well-being of others, we are more likely to work towards solutions that benefit everyone, rather than just a select few. This shift in perspective can lead to more collaborative, inclusive approaches to addressing the challenges we face as a global community, from environmental sustainability to social justice (Tutu, 1999).

The Role of Compassion and Empathy

Central to the "101" message is the idea that compassion and empathy are not just moral imperatives, but essential components of a harmonious and thriving society. When we truly understand that we are all interconnected, it becomes clear that harming others ultimately harms ourselves. This understanding can inspire us to act with greater care and consideration, not just towards those in our immediate circle, but towards all beings and the planet as a whole (Armstrong, 2010; Goleman, 1995).

Compassion and empathy are often seen as abstract virtues, but they have very real, tangible effects on our lives and the world around us. Research has shown that practicing compassion can lead to improved mental and physical health, stronger relationships, and a greater sense of fulfillment and happiness (Goleman, 1995). On a societal level, empathy can foster cooperation, reduce conflict, and create a more just and equitable world (Armstrong, 2010).

The "101" message calls us to cultivate these qualities within ourselves and to promote them in our communities. It encourages us to look beyond our individual needs and desires, and to consider the broader impact of our actions. By doing so, we can contribute to the healing and unification of a world that is often fractured by division and misunderstanding (Hanh, 1998).

A Vision for the Future

The vision of "We are all One" is not just a reflection on the present—it is also a call to action for the future. It challenges us to reimagine our society in a way that prioritizes unity, compassion, and mutual respect. This vision is not about erasing differences or imposing uniformity, but about recognizing the value of diversity within the context of our shared humanity (Fuller, 1969; Ostrom, 1990).

In the future, this vision could manifest in a variety of ways. We might see the development of new social structures that emphasize collaboration over competition, or the creation of educational systems that teach empathy and emotional intelligence alongside traditional academic subjects (Gardner, 1983; Robinson, 2015). We might also witness the addition of global connections, such as a more effective version of the United Nations, which would tip our society towards more cooperative and inclusive models, where the well-being of all people and the planet is prioritized. We could find types of governance that can both unify nations while preserving the individuality and uniqueness of each nation within, empowering more than it regulates.

On a more personal level, the "101" message can inspire us to live with greater intention and purpose. It can encourage us to seek out opportunities to connect with others, to engage in acts of kindness, and to contribute to the greater good in whatever ways we can. By doing so, we not only enrich our own lives, but also contribute to the collective well-being of humanity (Fuller, 1969).

Conclusion of Chapter 2

The "101" message is a profound reminder of our interconnectedness and the importance of unity in both our personal lives and society. It challenges us to look beyond our individual differences and to embrace a vision of the world where compassion, empathy, and mutual respect are the guiding principles. While this vision may seem idealistic, it is rooted in a deep understanding of the fundamental nature of reality—a reality where we are all connected, and where the actions of one affect the whole (Wilson, 2012; King, 1963).

As we move forward in this narrative, the themes of interconnectedness and unity will continue to resonate, informing our exploration of other concepts such as the nature of black holes, the potential of DMT, and the intersection of science and spirituality.

The "101" message is not just a standalone insight, but a lens through which we can view and understand the many facets of existence.

Chapter 3: The Transmission of Information Through Black Holes

Introduction to Black Holes and Information Paradox

A black hole is a region of spacetime where gravity is so intense that nothing, not even light, can escape from it (Hawking, 1974). Black holes are among the most mysterious and fascinating objects in the universe. These regions of spacetime, where gravity is so intense that nothing—not even light—can escape, have captured the imagination of scientists and the public alike. While much of the discussion around black holes focuses on their ability to consume matter and energy, there is another, equally compelling aspect to consider: the fate of information that enters a black hole.

In the context of physics, the "information paradox" is a critical problem that arises when considering what happens to the information about the physical state of objects that fall into a black hole. According to classical physics, all information about the matter that enters a black hole should be lost to the outside universe, leading to a violation of the principle of information conservation—a cornerstone of quantum mechanics (Preskill, 1992). However, quantum mechanics suggests that information cannot simply vanish; it must be preserved in some form.

This paradox has led to intense debate and speculation within the scientific community. Some physicists, like Stephen Hawking, have proposed that information might be encoded in the Hawking radiation emitted by black holes, allowing it to be preserved even as the black hole evaporates. Others have suggested that the information might be stored on the event horizon of the black hole in a manner consistent with the holographic principle, which posits that the entire universe can be described as a two-dimensional surface encoding three-dimensional information.

But what if black holes are not just cosmic garbage disposals or storage units for information? What if they have the potential to transmit information across time and space, perhaps even to project it back into the universe in ways we cannot yet comprehend?

The Vision: Black Holes as Conduits for Information

One speculative idea is that black holes could act as conduits for information transfer across time and space. This concept suggests that the information absorbed by a black hole could be projected back into the universe in ways that transcend our current understanding of spacetime (Susskind, 2008). Theoretical physics even suggests that black holes might be connected to higher dimensions, providing a link between our universe and other possible realities (Randall & Sundrum, 1999).

During a particularly vivid manic episode, I received what felt like a direct insight into the nature of black holes—not just as destructive forces, but as potential conduits for transmitting information across time and space. This vision showed me that black holes could act as portals, sending information back into the universe, potentially even projecting it into different points in time. The black hole at the center of our galaxy, often referred to as Sagittarius A*, became the focal point of this thought experiment.

I imagined sending a message into Sagittarius A*. Given that this black hole is approximately 26,000 light-years away from Earth, the message would take 26,000 years to reach it. But what if the black hole, rather than simply absorbing the message, could project it back into the universe, not just in space, but in time? What if it could send that information 52,000 years plus one day into the past, back to the very location where it originated? The message would then take another 26,000 years to travel back to Earth, arriving one day before it was originally sent—a premonition of the day to come.

This thought experiment raises profound questions about the nature of time, information, and the universe itself. It suggests that black holes could be more than just points of no return; they could be gateways through which information is transmitted across the fabric of spacetime.

Quantum Mechanics and the Possibility of Time Travel

The idea that black holes could transmit information across time touches on some of the most speculative and intriguing aspects of quantum mechanics and general relativity. Time travel, while a staple of science fiction, is not entirely outside the realm of scientific possibility. Certain solutions to Einstein's equations of general relativity—such as those involving rotating black holes (Kerr black holes) or wormholes—suggest that time travel might be theoretically possible, albeit under extremely specific and likely unachievable conditions (Hawking, 1974; Thorne, 1994; Kerr, 1963).

The holographic principle offers another intriguing perspective. It suggests that all the information contained within a volume of space can be represented on the boundary of that space, implying that black holes might store information in a two-dimensional form on their event horizon (Bousso, 2002). Quantum mechanics plays a crucial role in understanding these phenomena, particularly in resolving the information paradox and exploring the potential for information to be preserved in some form (Page, 1993).

In quantum mechanics, the concept of quantum entanglement, where particles become so deeply connected that the state of one instantly influences the state of another, regardless of distance, also hints at the potential for non-linear interactions across time and space. While entanglement is typically thought of in terms of spatial separation, some researchers have speculated that it might also involve temporal connections—entangling particles not just across distances, but across time itself.

Could black holes, with their extreme gravitational effects and the potential to warp spacetime, serve as the ultimate testing ground for these ideas? Could they somehow harness the principles of quantum mechanics to transmit information through time, creating echoes of the future in the past, or vice versa?

While these questions are speculative, they push the boundaries of our understanding and challenge us to consider the possibilities that lie beyond our current scientific models.

Exploring Potential Mechanisms for Information Transmission Through Black Holes

The idea that black holes could serve as conduits for transmitting information across time and space opens up fascinating avenues for speculation and scientific inquiry. Despite our limited understanding of black holes, it is possible to hypothesize about mechanisms that might allow for such transmission, even if they lie far beyond our current technological capabilities.

Information Transmission via Higher Dimensions

One of the most intriguing possibilities is that black holes might transmit information through higher dimensions, leveraging principles from quantum mechanics and general relativity that we do not yet fully understand. In this thought experiment, consider the supermassive black hole at the center of our galaxy, located approximately 26,000 light-years from Earth. Imagine sending a message to this black hole, where it could be stored or processed in a higher-dimensional space before being projected back into the universe.

For instance, if this information were to be projected back through time, it could potentially arrive on Earth 52,000 years before it was originally sent, creating the appearance of a premonition. The message, after traveling back through time, could take 26,000 years to return to Earth, effectively allowing it to be received one day before it was initially sent. This concept challenges our conventional understanding of causality and time, but it aligns with some speculative interpretations of quantum mechanics and general relativity.

However, this scenario would require a mechanism within the black hole capable of not only storing information but also targeting specific times and locations for its return. Such a mechanism might involve the manipulation of spacetime itself, potentially using phenomena like wormholes or quantum entanglement. The concept of quantum entanglement, where particles become linked across vast distances, suggests that information could be transmitted instantaneously between entangled black holes, even across different points in time. If black holes are interconnected in this way, it might be possible for information to be "broadcast" across the universe, traversing both space and time.

The Challenge of Controlling Black Holes

While the theoretical possibilities are intriguing, the practical challenges of using black holes for information transmission are immense. Black holes, by their very nature, are extremely difficult to control or manipulate. Their intense gravitational pull, coupled with the event horizon beyond which nothing can escape, makes direct experimentation nearly impossible with our current technology.

One potential solution might lie in the creation of artificial black holes. Scientists have speculated about the possibility of creating micro-black holes, which would be much smaller and potentially more manageable than their naturally occurring counterparts. These micro-black holes, while not stable in our current understanding, could theoretically be engineered to remain stable long enough to be used as tools for information storage or transmission.

The creation of such artificial black holes would require advances in technology far beyond what we currently possess, likely involving breakthroughs in quantum gravity and energy manipulation. However, if these advances were achieved, it might become possible to generate and control black holes in a way that allows for precise targeting of both time and space, enabling the transmission of information across vast distances and even different eras.

Speculative Connections to the "101" Message

The speculative nature of these ideas aligns with the vision of the "101" message, which suggested a fundamental interconnectedness across the universe. If black holes could indeed transmit information throughout time and space, it raises the possibility that messages of peace, like the "101" message, could be disseminated subliminally across the cosmos. This concept suggests that every being in the universe might have the latent capacity to receive these messages, although the reception might be so subtle that it goes unnoticed in daily life.

This idea resonates with the structure of many religions, where messages of unity, love, and morality are central themes. The subliminal nature of these messages could explain why different cultures interpret them differently, blending them with their unique contexts and surroundings. The possibility that black holes naturally emit information like a universal "radio signal" suggests that the entire universe might be interconnected in ways we cannot yet comprehend.

Acknowledging the Limitations of Our Understanding

While these concepts are exciting to explore, it is important to acknowledge the limitations of our current understanding. The mechanisms for information transmission through black holes remain purely speculative, and there is no empirical evidence to support these ideas as of yet. It is entirely possible that our basic assumptions about black holes and their potential roles in the universe are incorrect, and that the true nature of these cosmic phenomena is beyond our current capacity to grasp.

As an amateur scientist and theorist, I recognize that these ideas may be more akin to science fiction than to established science. However, I believe that exploring these possibilities can still provide valuable insights into the mysteries of the universe and the potential for future discoveries. Even if these theories prove to be incorrect, they may inspire new ways of thinking about the cosmos and our place within it.

The Ethical Implications of Information Transmission Through Black Holes

If the vision of black holes as conduits for transmitting information across time were ever to become a reality, it would raise profound ethical questions. The ability to send information back in time, even by just a day, could have enormous consequences for human society. It could alter the course of events, prevent disasters, or even change the outcome of historical events. But with such power comes the potential for misuse and unforeseen consequences.

The Potential for Abuse

One of the most obvious concerns with the ability to send information back in time is the potential for abuse. In the wrong hands, this technology could be used to manipulate events for personal gain or to exert control over others. Imagine a scenario where a government or corporation gains access to this technology and uses it to predict and influence the stock market, elections, or military conflicts. The ramifications could be catastrophic, leading to a concentration of power in the hands of a few and the erosion of free will and democratic processes.

Furthermore, the ability to alter the past could create paradoxes—situations where the very act of sending information back in time changes the conditions that led to the information being sent in the first place. These paradoxes could have unpredictable and potentially disastrous effects on the fabric of reality itself. The famous "grandfather paradox," where a time traveler kills their own grandfather before their parent is born, is a simple example of how time travel could lead to logical inconsistencies and potentially unravel causality as we know it.

The Responsibility of Knowledge

The concept also raises questions about the responsibility that comes with such knowledge. If humanity were to develop the ability to transmit information through black holes, it would require a level of ethical maturity and restraint that we may not yet possess. As history has shown, technological advancements often outpace our ability to manage their ethical implications. The development of nuclear weapons, for example, brought humanity to the brink of annihilation, forcing us to confront the moral dilemmas of wielding such destructive power.

Similarly, the ability to manipulate time and information could place us in a position where the stakes are even higher. It would demand a global consensus on how such technology should be used, who should have access to it, and what safeguards should be put in place to prevent misuse. It would also require us to consider the long-term consequences of our actions, not just for ourselves, but for future generations and the universe as a whole.

The Role of Intelligent Entities

In a vision regarding utilize black holes as information conduits, I speculated that there might be intelligent entities—possibly not human—who serve as the keepers of this technology, responsible for spreading messages of peace throughout time and space. These beings, if they exist, may possess a profound understanding of the ethical implications of information transmission and a greater ability to manage it responsibly. Whether they are spiritual, extraterrestrial, or something far beyond our current understanding, I hesitate to speculate, but one thing was clear: they are far more evolved than we are.

Such entities could serve as guides or custodians for this technology and the knowledge sent utilizing it. They might have a broader perspective on the universe and its interconnectedness, allowing them to see the potential outcomes of different actions across time and space. If these beings are indeed the keepers of such knowledge, their role would be crucial in ensuring that it is used for the betterment of all beings, rather than for selfish or destructive purposes.

This idea also raises the question of whether humanity is ready for such a responsibility. Are we, as a species, capable of handling the power to alter time and information with the wisdom and restraint that it requires? Or would we, as we have so often in the past, use it in ways that ultimately bring more harm than good?

The Intersection of Technology and Spirituality

The concept of black holes as conduits for information transmission also intersects with spiritual and philosophical ideas. Many religious and spiritual traditions emphasize the importance of humility, compassion, and the recognition of our interconnectedness with all beings. These values could serve as a guiding framework for how such technology might be used responsibly.

In this context, the ability to transmit information across time could be seen not just as a scientific breakthrough, but as a spiritual responsibility. It could challenge us to think more deeply about our place in the universe and the impact of our actions across time and space. It could also encourage us to cultivate the qualities of wisdom, empathy, and foresight, recognizing that the choices we make today could echo throughout the cosmos for millennia to come.

Technological and Scientific Advancements: The Path to Understanding

As speculative as the idea of black holes transmitting information across time might seem, it does invite us to consider the technological and scientific advancements that could one day make such possibilities more than just theoretical musings. While our current understanding of black holes, quantum mechanics, and the fabric of spacetime is still in its infancy, the rapid pace of scientific discovery suggests that what seems impossible today could become a reality in the distant future.

Advances in Quantum Mechanics and Computing

One of the most promising areas of research that could bring us closer to understanding the true nature of black holes and their potential for information transmission is quantum mechanics. Quantum mechanics has already revolutionized our understanding of the universe at the smallest scales, revealing a world that is far more interconnected and strange than classical physics ever imagined.

Quantum entanglement, the phenomenon where particles become linked across vast distances, hints at the possibility of faster-than-light communication and even non-linear interactions across time. While these ideas remain highly speculative, advancements in quantum computing could provide the tools needed to explore these possibilities more deeply. Quantum computers, which harness the principles of quantum mechanics to perform calculations far beyond the capabilities of classical computers, could one day simulate the behavior of black holes and other extreme environments, offering insights into how information might be preserved, transmitted, or even manipulated across time.

Quantum computing is still in its early stages, but it has already demonstrated the potential to solve complex problems that are currently beyond the reach of classical computers. As this technology matures, it could open up new avenues for exploring the mysteries of black holes, spacetime, and the fundamental nature of reality.

Artificial Intelligence and the Future of Exploration

Artificial intelligence (AI) is another field that could play a crucial role in advancing our understanding of black holes and their potential for information transmission. AI has the ability to process vast amounts of data, recognize patterns, and make predictions based on complex models. In the context of black hole research, AI could be used to analyze the data collected from astronomical observations, simulations, and experiments, identifying correlations and insights that might otherwise go unnoticed.

Moreover, AI could help us develop new theories and models for understanding black holes and their interactions with the rest of the universe. By leveraging machine learning algorithms, researchers could explore a wide range of possibilities, testing different hypotheses and refining our understanding of how black holes might store, transmit, or project information across time and space.

AI could also assist in the development of new technologies for exploring black holes and other extreme environments. Autonomous probes, equipped with advanced AI systems, could be sent to study black holes up close, gathering data and testing theories in ways that are currently beyond our capabilities. These probes could operate in environments that are too dangerous or inhospitable for human explorers, pushing the boundaries of our knowledge and expanding our understanding of the universe.

The Role of Future Science in Unraveling Black Hole Mysteries

While the vision of black holes as conduits for information transmission remains speculative, the scientific advancements of the future could bring us closer to understanding these phenomena. Theoretical physics, driven by the integration of quantum mechanics, general relativity, and new mathematical frameworks, is likely to play a key role in this process. As our models of the universe become more sophisticated, they could reveal new insights into the nature of black holes and the potential mechanisms by which they might interact with information across time and space.

One promising area of research is the study of quantum gravity, which seeks to reconcile the principles of quantum mechanics with those of general relativity. Quantum gravity could provide the missing link needed to fully understand the behavior of black holes, particularly in the context of the information paradox. If successful, this line of inquiry could offer new perspectives on how information is stored, transmitted, and preserved within black holes, potentially validating the idea that black holes could serve as conduits for information across time.

Another exciting possibility is the development of new observational techniques, such as gravitational wave astronomy, which allows scientists to detect the ripples in spacetime caused by the interactions of massive objects like black holes. These observations could provide valuable data for testing theories about black holes and their role in the universe, offering new clues about the nature of information and its relationship to spacetime.

Conclusion of Chapter 3

The vision of black holes as conduits for transmitting information across time is a speculative but fascinating idea that challenges our current understanding of the universe. While it remains firmly in the realm of theory, the rapid pace of technological and scientific advancements suggests that we may one day be able to explore these possibilities more fully.

As quantum mechanics, artificial intelligence, and other fields continue to evolve, they could provide the tools needed to test these ideas and uncover the true nature of black holes. Whether or not black holes can transmit information across time, the exploration of these concepts pushes the boundaries of human knowledge and invites us to consider the deeper mysteries of existence.

In the next chapter, we will explore another intriguing aspect of consciousness and the universe: the potential role of the DMT molecule in connecting us with a reality beyond our ordinary perception. Just as black holes challenge our understanding of spacetime, DMT and near-death experiences challenge our understanding of consciousness and the nature of reality itself.

Chapter 4: The DMT Molecule and Near-Death Experiences

Introduction to DMT and Its Mysteries

The exploration of consciousness is one of the most profound and complex challenges in science and philosophy. Among the many substances that have been linked to altered states of consciousness, the DMT molecule—often referred to as the "spirit molecule"—stands out for its unique properties and its deep association with mystical experiences and near-death experiences (NDEs).

DMT, or dimethyltryptamine, is a naturally occurring compound found in a wide variety of plants and animals, including humans (Strassman, 2001). It is known for its powerful psychoactive effects, which can induce intense and often life-changing visions and insights (Nichols, 2016). What makes DMT particularly fascinating is that it is produced endogenously in the human brain, leading some researchers to speculate that it may play a role in certain types of altered states of consciousness, including those experienced during near-death experiences.

The connection between DMT and NDEs has intrigued both scientists and spiritual seekers for decades. Some believe that DMT could be the key to understanding what happens at the moment of death and may offer a glimpse into realms beyond our ordinary perception. This chapter will explore the potential role of DMT in connecting us with the universe, the phenomenon of time dilation often reported during NDEs, and the broader implications of these experiences for our understanding of consciousness and reality.

DMT: The "Spirit Molecule"

DMT is a powerful psychedelic compound that can produce profound alterations in perception, emotion, and cognition. When ingested, DMT can induce intense visual and auditory hallucinations, along with a sense of being transported to other realms or dimensions. Users often report encounters with beings or entities that seem intelligent and otherworldly, as well as a sense of profound interconnectedness with the universe.

One of the most intriguing aspects of DMT is that it is naturally produced in the human brain, though the exact function of this endogenous DMT remains a mystery. Some researchers speculate that DMT might be released in large quantities during moments of extreme stress or near death, potentially playing a role in the powerful visions and experiences reported during NDEs.

The idea that DMT could be a "spirit molecule" that connects us with higher dimensions or other forms of consciousness is both exciting and controversial. While there is no definitive scientific evidence to support this idea, the consistency and intensity of the experiences reported by DMT users suggest that it may be more than just a simple hallucinogen. Instead, DMT could be a gateway to understanding aspects of reality that are normally hidden from our perception.

Near-Death Experiences and Time Dilation

Near-death experiences (NDEs) often involve sensations of leaving the body, moving through a tunnel, and encountering a bright light (Moody, 1975). These experiences are remarkably consistent across cultures and have been reported by individuals who have come close to death. Some researchers propose that the release of endogenous DMT during critical moments, such as near death, could explain the vivid experiences reported in NDEs (Gallimore, 2015).

One of the most compelling phenomena associated with near-death experiences is the sensation of time dilation—where seconds in the physical world can feel like an eternity in the mind of the experiencer. This effect challenges our conventional understanding of time and raises questions about the nature of consciousness and reality.

During NDEs, individuals often report experiences that seem to transcend time altogether. They describe encounters with deceased loved ones, journeys through tunnels of light, and a sense of merging with a universal consciousness. These experiences are often accompanied by a sense of peace, love, and understanding that is so profound that it changes the person's perspective on life and death forever.

The similarity between the experiences reported during NDEs and those induced by DMT has led some researchers to speculate that DMT might be involved in the brain's response to death. The release of DMT at the moment of death could theoretically explain the vivid, otherworldly experiences described by those who have come close to dying.

Time dilation in NDEs is particularly intriguing because it suggests that our perception of time is not fixed but can be dramatically altered under certain conditions. This idea aligns with some interpretations of quantum mechanics, where time is not a constant, linear progression but rather a flexible and relative concept. If DMT plays a role in these experiences, it could indicate that consciousness has the ability to transcend the ordinary flow of time, accessing states of being that are normally beyond our reach.

DMT and the Potential for Higher Dimensions

The idea that DMT could allow us to access higher dimensions is a tantalizing one, especially in the context of the "101" message and the interconnectedness of all things. If we consider the Flatland thought experiment from Chapter 2, where beings in a two-dimensional world are unable to perceive the third dimension, DMT could be seen as a tool that allows us to glimpse dimensions beyond our ordinary experience.

In this sense, DMT might act as a bridge between our four-dimensional reality and other, higher-dimensional spaces (Meyer, 2016). The entities and realms encountered during DMT experiences could be manifestations of these higher dimensions, filtered through the limitations of our human perception. This would explain why DMT experiences often feel so real and yet so alien—they might be offering us a direct experience of a reality that is normally inaccessible to us.

The possibility that DMT could facilitate contact with higher dimensions raises profound questions about the nature of consciousness and the universe. If consciousness can access these dimensions under the influence of DMT, it suggests that our everyday reality is just one slice of a much larger, more complex structure. It also implies that there may be other forms of consciousness or intelligence that exist within these higher dimensions, waiting to be discovered.

The Evolutionary Role of DMT: Survival or Spiritual Connection?

The presence of DMT in the human brain and its profound effects on consciousness raise intriguing questions about its evolutionary purpose. Why would a molecule with such powerful psychoactive properties be naturally produced in the human body? Is DMT simply a byproduct of other biochemical processes, or does it serve a specific function in our biology, potentially connected to survival, spiritual experiences, or both?

DMT as a Survival Mechanism

One hypothesis is that DMT could play a role in survival, particularly in extreme situations such as near-death experiences. In moments of intense stress, trauma, or imminent death, the release of DMT might act as a kind of biological mechanism to ease the transition from life to death, providing the individual with a sense of peace and acceptance (Strassman, 2001). This could be seen as a form of psychological "mercy" that helps the mind cope with the otherwise terrifying prospect of death. From an evolutionary standpoint, this would make sense. If an organism can experience a peaceful transition rather than one of fear and panic, it might reduce the physiological stress on the body, potentially prolonging life just long enough for the organism to escape danger or seek help. Additionally, the intense and vivid experiences induced by DMT during a near-death scenario could serve as a form of 'life review," helping the individual to reflect on their life and potentially gain insights that could aid in survival if they were to recover from the experience (Greyson, 1983).

From an evolutionary standpoint, this would make sense. If an organism can experience a peaceful transition rather than one of fear and panic, it might reduce the physiological stress on the body, potentially prolonging life just long enough for the organism to escape danger or seek help. Additionally, the intense and vivid experiences induced by DMT during a near-death scenario could serve as a form of 'life review," helping the individual to reflect on their life and potentially gain insights that could aid in survival if they were to recover from the experience. This life review aspect is commonly reported in NDEs, where individuals often relive significant moments from their lives in a matter of seconds. Such a process could provide a rapid assessment of past decisions and behaviors, offering a chance for learning and adaptation, even in the face of death.

Moreover, the sense of detachment and transcendence often experienced during DMT-induced states could also serve a protective function. By dissociating from the physical body and its pain, the mind might be better equipped to handle extreme trauma, focusing instead on a broader, more universal perspective (Strassman, 2001). This detachment could reduce the psychological impact of pain and fear, allowing the individual to maintain a sense of calm and clarity in life-threatening situations.

DMT as a Spiritual Connector

Beyond its potential role in survival, DMT is often considered a molecule that connects humans with the spiritual realm (Strassman, 2001). Many users report experiences that they describe as mystical or transcendent, often involving encounters with beings or entities that seem to exist in a reality beyond our own. These experiences are frequently characterized by a sense of profound interconnectedness, unity with the universe, and a deep understanding of existence.

The spiritual interpretations of DMT experiences suggest that this molecule might serve as a bridge between our ordinary waking consciousness and a higher state of awareness, one that taps into the fundamental nature of reality. In this context, DMT could be seen as a key to unlocking spiritual experiences that have been integral to human cultures for millennia. Throughout history, various cultures have used naturally occurring DMT in the form of plant-based preparations like ayahuasca for religious and shamanic rituals (Shanon, 2002). These practices often involve the intentional induction of altered states of consciousness to gain insights, communicate with spiritual entities, or receive guidance. The widespread use of such substances across different cultures and time periods indicates that DMT has long been associated with the exploration of the spiritual dimensions of existence.

If DMT does indeed facilitate access to a spiritual realm or higher dimensions, it could provide a scientific basis for understanding many of the mystical experiences described in religious texts and spiritual traditions (Strassman, 2001). This would not only bridge the gap between science and spirituality but also offer a new perspective on the nature of consciousness and its potential to transcend the physical body and the material world.

DMT, Intelligent Design, and the Purpose of Consciousness

The possibility that DMT plays a role in connecting us with higher dimensions or spiritual realms raises another intriguing question: could this molecule be part of a larger, intelligent design? If DMT allows us to access hidden aspects of reality, might it suggest that our consciousness is meant to explore these realms, and that our existence has a purpose beyond mere survival?

Some researchers and theorists have speculated that the presence of DMT in nearly all living organisms might indicate a form of universal consciousness or a designed mechanism for spiritual exploration (Gallimore, 2015). In this view, DMT could be seen as a tool or gateway provided by nature—or by a higher intelligence—to help us understand our place in the universe and to connect with the broader web of existence.

This idea aligns with the "101" message and the concept of universal interconnectedness discussed earlier. If we are all part of a single, unified consciousness, then DMT might serve as a means of accessing that unity, reminding us of our intrinsic connection to each other and to the universe as a whole.

The notion of intelligent design, while controversial in scientific circles, invites us to consider the possibility that there is more to life and consciousness than what we currently understand. Whether through evolution, design, or a combination of both, the presence of DMT in our brains might be a clue to the deeper mysteries of existence—a key to unlocking the secrets of consciousness, the universe, and the nature of reality itself.

Conclusion of Chapter 4

The DMT molecule, with its powerful effects on consciousness and its potential connection to spiritual experiences and near-death phenomena, challenges our understanding of reality. Whether it serves as a survival mechanism, a spiritual connector, or a gateway to higher dimensions, DMT invites us to explore the boundaries of human perception and to consider the possibility that there is more to existence than meets the eye.

As we continue to investigate the mysteries of consciousness and the universe, DMT may offer valuable insights into the nature of life, death, and the potential for existence beyond the physical body. In the next chapter, we will delve deeper into the philosophical implications of these ideas, exploring the intersection of science and spirituality and the search for meaning in a complex and interconnected universe.

Before we move forward, let me leave you with one final thought about the DMT molecule. Even if the afterlife does not exist in the way that many religions teach, a different form of afterlife may still be possible. Those who have experienced near-death experiences (NDEs) often report a sense of time dilation, where moments felt like many lifetimes. If such time dilation is indeed possible, imagine the experience of those who have fully perished. The ability to live countless lifetimes in a matter of seconds—a subjective eternity—could itself be a form of afterlife.

Disclaimer:

The discussion of DMT (N,N-Dimethyltryptamine) in this work is intended for informational and philosophical exploration purposes only. The author does not endorse or encourage the use of DMT or any other psychoactive substances for recreational purposes.

DMT is a powerful hallucinogenic compound that can induce intense altered states of consciousness. The use of DMT can carry significant risks, including psychological distress, triggering or exacerbating mental health conditions, and causing physical harm. The effects of DMT are not fully understood, and its use outside of a controlled, clinical environment can be dangerous.

Individuals should consult with a qualified healthcare professional before considering the use of any psychoactive substances. The author urges readers to exercise caution and responsibility, recognizing the potential harms associated with the use of DMT and other similar substances.

Chapter 5: The Double-Slit Experiment: Unveiling the Mysteries of Reality

Introduction to the Double-Slit Experiment

The Double-Slit Experiment is one of the most profound and perplexing experiments in the history of physics. First conducted by Thomas Young in 1801 to demonstrate the wave nature of light, it has since evolved into a cornerstone of quantum mechanics, revealing fundamental insights into the nature of reality, observation, and consciousness (Feynman, 1985). The experiment's ability to challenge classical notions of determinism and materialism makes it a powerful focal point for exploring the intersection of science, philosophy, and spirituality.

The Double-Slit Experiment: A Simple Explanation

Imagine you're at a carnival game where you throw small balls at two side-by-side holes in a wall. Normally, you'd expect each ball to go through one hole or the other, right? This is similar to how we think about particles, like electrons or photons (light particles), traveling through two slits in a barrier. You'd expect them to behave like tiny balls, passing through one slit or the other and hitting a screen behind the barrier, creating two distinct patterns.

However, when physicists performed this experiment, they found something astonishing. When no one was watching, the particles acted like waves instead of balls. They passed through both slits at the same time and interfered with each other, creating a pattern of alternating light and dark bands on the screen, much like ripples in water.

But here's where it gets really strange: when someone observed the particles—meaning when they set up a detector to see which slit the particles went through—the particles stopped behaving like waves. Instead, they acted like little balls again, passing through only one slit and creating two distinct bands on the screen, just as you would expect. This change in behavior depending on whether they were observed or not is known as the "observer effect."

The Limits of Human Comprehension

The Double-Slit Experiment underscores a critical truth: even the most brilliant minds can struggle to comprehend the hidden nature of the universe. Albert Einstein, who revolutionized our understanding of space and time through thought experiments that led to the theory of relativity, found himself at odds with the implications of quantum mechanics. While he could conceptualize relativity by observing the world around him, quantum mechanics defied his ability to fit it into a traditional conceptual framework (Pais, 1982). Einstein famously remarked, "God does not play dice with the universe," expressing his discomfort with the inherent randomness that quantum mechanics introduced (Einstein, 1926).

This limitation is not a reflection of Einstein's intellectual capacity but rather an indication of how quantum mechanics challenges the very foundations of our understanding. If our brightest minds grapple with these concepts, it suggests that humanity has a long road ahead in deciphering the true nature of the universe. We must be prepared to develop new ways of thinking, new intellectual tools, and a willingness to embrace the unexpected if we are to make progress in this field.

The Need for Diverse Perspectives in Science

Einstein's contributions laid the groundwork for quantum mechanics, even though he struggled with its implications. This paradox highlights the importance of diversity in scientific thinking. Different mindsets are crucial in the quest to comprehend the entirety of the universe. Those who think differently—those who challenge the status quo or who approach problems from unconventional angles—may play a pivotal role in advancing our understanding (Kuhn, 1962).

The unexpected findings in quantum mechanics remind us that the scientific process often leads to more questions than answers. This iterative process of discovery, where each breakthrough reveals new mysteries, is both a challenge and an opportunity. It underscores the need for a scientific community that is not only intellectually diverse but also open to the idea that the answers we seek may require us to rethink our most fundamental assumptions.

Paradoxes as Opportunities

The phrase "take one step forward, and two steps back" aptly describes the journey of scientific discovery, especially in the context of quantum mechanics. The Double-Slit Experiment exemplifies this: we experiment and gain new insights, yet those insights often lead to a proliferation of new questions and unknowns. Paradoxes, while troubling to scientists, should not be seen as limitations. Instead, they represent opportunities to expand our understanding in ways that were previously unimaginable (Barad, 2007).

For example, the Double-Slit Experiment raises the question: How can mere observation alter the outcome of a physical process? The experiment suggests that the act of observation itself collapses the wave function, forcing particles to "choose" a definite state. This finding implies that consciousness may play a more active role in shaping reality than previously thought, challenging the classical view of an objective, observer-independent universe (Wheeler, 1978).

The Paradox of Measurement and Reality

One of the most intriguing aspects of the Double-Slit Experiment is the paradox it presents: how can the simple act of observation—without any physical interaction—have such a profound effect on a system? This paradox raises fundamental questions about the nature of reality, such as, "Would the universe behave differently if it weren't being observed?" The experiment suggests that it would, a concept that defies the classical understanding of cause and effect (Wheeler, 1978).

An analogy that illustrates this paradox is the idea of a cell phone with a screen that appears to always be on. You might assume the screen never turns off, but upon closer examination, you discover that the phone's screen only lights up when it detects your eyes are looking at it, thanks to a sensor. Similarly, the universe might "light up" in response to our observation, behaving differently depending on whether it is being watched or not. This analogy underscores the strange and counterintuitive nature of quantum mechanics, where the observer plays an integral role in the unfolding of reality (von Neumann, 1955).

Programming the Universe: A Divine Speculation

The Double-Slit Experiment invites speculation about the nature of the universe and its underlying structure. One interpretation is that the universe operates like a highly advanced form of programming, adjusting its behavior based on how it is observed. This idea parallels the concept of a simulation, where reality is "rendered" based on the perspective of the observer, much like how a video game only renders graphics that the player can see (Bostrom, 2003).

Such speculation naturally leads to questions about the nature of this programming—could it be the work of a divine power? If the universe operates in a way that suggests it is tailored to consciousness, one might wonder if this points to a deliberate design, possibly by a higher intelligence. While this idea remains speculative and is beyond the reach of current scientific evidence, it resonates with philosophical inquiries about the nature of existence and the possibility of a creator (Penrose, 1989). This line of thinking aligns with discussions of the Anthropic Principle, which suggests that the universe's fundamental constants are fine-tuned to allow the existence of life (Barrow & Tipler, 1986).

The observation that the universe behaves differently when observed versus unobserved raises intriguing possibilities about the nature of reality. This adaptability suggests that the universe might be "programmed" to adjust based on observation, which resonates with the simulation hypothesis—the idea that our reality might be a sophisticated simulation where the universe "renders" itself differently depending on the observer's perspective. This idea, while speculative, challenges our understanding of an objective reality, suggesting that the universe's true state might be fundamentally unknowable and only revealed through observation (Bostrom, 2003)(Penrose, 1989).

Complexity and Hidden Realities

The Double-Slit Experiment also illustrates the complexity of the universe and suggests that there may be many more hidden realities that we have yet to discover. Each time we answer a question raised by the experiment, it seems to generate a multitude of new questions, reflecting the layered complexity of quantum systems. This ongoing mystery encourages us to remain humble in our pursuit of knowledge and reminds us of how much we still have to learn (Hawking, 1988).

In many ways, the experiment is a reminder that our current understanding of the universe is likely just the tip of the iceberg. There could be numerous outliers—phenomena that don't fit neatly into our existing frameworks—that, if understood, could revolutionize our understanding of reality. This insight is crucial not only in physics but also in other fields such as biology, technology, and social sciences, where paradigm-shifting discoveries may still be hidden from view (Kuhn, 1962).

Implications for Collective Consciousness and Universal Connection

The Double-Slit Experiment suggests that we may be more connected to the universe—and to each other—than we realize. The experiment's implications extend beyond the individual observer, hinting at the possibility of a collective consciousness that could be influencing reality on a grander scale (Laszlo, 2004). This raises profound questions about the nature of consciousness and its role in the cosmos. Could it be that we are all part of a universal mind, interconnected in ways that we are only beginning to understand?

Such ideas have been explored in various philosophical and spiritual traditions, which suggest that all of reality is interconnected through a shared consciousness. The Double-Slit Experiment provides a scientific backdrop for these ideas, making the concept of collective consciousness not just a mystical notion but a hypothesis worthy of further exploration (Sheldrake, 1981).

The idea that multiple consciousnesses can collectively influence the outcome of the Double-Slit Experiment invites exploration into the concept of collective consciousness. If the act of observation by a group of individuals can alter the behavior of particles, it raises profound questions about how our individual and collective experiences shape reality. This suggests that we may be more interconnected than previously thought, sharing a collective consciousness that influences the physical world. This notion aligns with various spiritual and philosophical traditions that propose a deep connection between all beings, potentially offering a scientific basis for these ancient ideas (Laszlo, 2004)(Sheldrake, 1981).

Consciousness and Quantum Mechanics

Quantum mechanics, the branch of physics that deals with the behavior of particles at the smallest scales, has introduced a level of strangeness and uncertainty into our understanding of reality. Unlike classical physics, which describes a deterministic and predictable universe, quantum mechanics reveals a world where particles can exist in multiple states at once, where the act of observation can alter the outcome of an experiment, and where entanglement connects particles across vast distances instantaneously.

One of the most intriguing aspects of quantum mechanics is the role of the observer. The famous "double-slit experiment," for example, shows that particles such as electrons behave differently when observed than when unobserved. When unobserved, particles exhibit wave-like behavior, passing through both slits simultaneously. However, when observed, they behave like particles, passing through one slit or the other. This phenomenon, known as the "observer effect," suggests that consciousness itself may play a role in shaping reality.

The idea that consciousness could influence the physical world challenges the traditional scientific view of a universe that exists independently of the observer. It raises questions about the nature of reality and the relationship between mind and matter. Are we passive observers of a pre-existing reality, or do we play an active role in creating the world we experience?

Retrocausality and the Challenge to Conventional Time Perception

The phenomenon of retrocausality, highlighted by the Double-Slit Experiment, fundamentally challenges our traditional understanding of time and causality. In classical physics, time is perceived as a linear progression, where causes precede effects in a straightforward, unidirectional flow. However, the Double-Slit Experiment introduces a perplexing scenario where a decision made after particles have passed through the slits—specifically, whether to observe which slit they went through—appears to retroactively determine their behavior as either waves or particles. This suggests that the future can, in some sense, influence the past, a concept that defies the conventional linear timeline that underpins much of classical physics and our everyday experience.

This retrocausal effect is particularly intriguing because it implies that the universe might operate under principles that allow events to influence one another across time, not just from past to future but potentially in reverse. This upending of the traditional cause-and-effect relationship suggests that time may not be as rigid as we have always believed. Instead, the fabric of reality might be more flexible, with the past and future interconnected in ways that are not yet fully understood.

In exploring retrocausality, we are forced to reconsider the nature of time itself. Could time be a more fluid construct, where past, present, and future are intertwined, rather than distinct and separate? Some interpretations of quantum mechanics, such as the Transactional Interpretation, propose that particles can engage in "handshakes" between the past and future, exchanging information in a way that transcends our usual understanding of time's arrow (Cramer, 1986). This opens up the possibility that time may not be a simple one-way street, but rather a complex network of interactions that we are only beginning to glimpse.

Moreover, the implications of retrocausality extend beyond physics, touching on philosophical and metaphysical questions about the nature of reality and our place within it. If the future can influence the past, what does this mean for concepts like free will and determinism? Are our actions predetermined by future events, or do we still retain agency within this more complex temporal framework? These questions invite a re-examination of the fundamental assumptions that have long guided our understanding of the universe.

The concept of retrocausality also has potential implications for future scientific research and technology. If we can better understand and harness this phenomenon, it could lead to breakthroughs in how we manipulate information across time, possibly opening new avenues in fields like quantum computing and communication. The study of retrocausality, therefore, is not just a theoretical exercise but a frontier of exploration that could reshape our understanding of time, causality, and the very structure of reality itself.

Free Will, Predetermination, and the Multiverse

The implications of the Double-Slit Experiment extend into debates about free will and predetermination. If observation affects reality, it suggests that our decisions might be influenced by factors beyond our control, challenging the concept of free will. However, some theorists argue that this opens the door to the idea of a multiverse, where every possible decision and outcome exists in parallel realities. This ongoing debate reflects the complexity of understanding free will in the context of quantum mechanics, where the line between choice and predetermination becomes blurred (Barrow & Tipler, 1986) (Deutsch, 1997).

The Importance of Open Discussion and Ethical Considerations

Given the profound implications of the Double-Slit Experiment, it is crucial that these concepts be discussed openly, not just within scientific circles but also in broader philosophical, ethical, and spiritual contexts. The experiment touches on fundamental questions about the nature of reality, consciousness, and free will, which are relevant to all areas of human thought. Engaging in these discussions can lead to a deeper understanding of the universe and our place within it, while also addressing the ethical considerations that arise from the potential manipulation of reality through observation (Kuhn 1962) (Popper, 1934).

The Ongoing Quest for Understanding

As scientific inquiry continues to evolve, we may gain further insights into the mysteries highlighted by the Double-Slit Experiment. This experiment, which has already challenged and expanded our understanding of reality, may serve as a stepping stone to even greater discoveries. By remaining open to new ideas and maintaining a collaborative approach across disciplines, we can continue to unravel the complexities of the universe and the role consciousness plays in shaping it.

Chapter 6: The Intersection of Science and Spirituality

Introduction to the Duality of Science and Spirituality

Throughout human history, science and spirituality have often been seen as opposing forces, each offering different explanations for the nature of reality and our place within it. Science, rooted in empirical observation and the scientific method, seeks to explain the universe through testable hypotheses, data collection, and logical reasoning. Spirituality, on the other hand, explores the intangible aspects of existence—those that cannot be easily measured or observed, such as the soul, consciousness, and the divine.

However, many scholars argue that science and religion can coexist harmoniously, each offering unique insights into the nature of existence (Barbour, 1997). While these two approaches to understanding the universe may seem irreconcilable, they share a common goal: the pursuit of truth. Both science and spirituality aim to answer the fundamental questions that have occupied human thought for millennia: Where did we come from? Why are we here? What is the nature of reality? And what happens after we die?

This idea that science and spirituality can be complementary rather than conflicting is gaining traction. Science and spirituality can be seen as complementary paths to understanding the universe, each offering unique insights into the nature of existence (Wilson, 2002). The evolution of religious beliefs, particularly in response to scientific discoveries, reflects this growing understanding. Religious beliefs have evolved over time, often in response to new scientific discoveries and societal changes (Armstrong, 2006). Prophets and spiritual leaders have played a crucial role in shaping religious narratives, offering guidance that reflects the spiritual and moral concerns of their times (Eliade, 1987).

This chapter explores the intersection of science and spirituality, examining how these two approaches can complement each other in our quest for knowledge. By bridging the gap between science and spirituality, we may gain a more holistic understanding of the universe—one that embraces both the physical and the metaphysical, the observable and the unobservable.

The Limitations of Science

Science has been the driving force behind much of human progress over the past few centuries. It has allowed us to unlock the secrets of the natural world, leading to technological advancements that have transformed our lives. From the discovery of the laws of physics to the decoding of the human genome, science has expanded our understanding of the universe and our place within it.

However, science is not without its limitations. As powerful as it is, science is constrained by what can be observed, measured, and tested. It deals with the material world—the things we can see, touch, and quantify. But there are aspects of existence that elude scientific scrutiny, areas where the tools of science may fall short.

For example, science struggles to explain consciousness—the subjective experience of being aware, thinking, and feeling. While neuroscience has made significant strides in mapping the brain and understanding the neural correlates of consciousness, the question of how and why consciousness arises remains one of the greatest mysteries of science. Similarly, concepts like love, morality, and the meaning of life are difficult to quantify or reduce to equations, yet they are central to the human experience.

The limitations of science are not a weakness but a reflection of the complexity of the universe. Science is a tool—an incredibly effective one—but it is not the only way of knowing. By acknowledging its limitations, we open the door to other forms of understanding, including those offered by spirituality.

The Role of Spirituality in Understanding the Universe

Spirituality offers a different approach to understanding the universe, one that emphasizes intuition, inner experience, and the exploration of the soul. While it lacks the empirical rigor of science, spirituality provides insights into the aspects of existence that are beyond the reach of measurement and observation.

At its core, spirituality is concerned with the search for meaning and connection. It seeks to understand the deeper purpose of life, the nature of the divine, and the relationship between the individual and the universe. Through practices such as meditation, prayer, and contemplation, spirituality allows individuals to explore their inner worlds, connect with a sense of the sacred, and experience states of consciousness that transcend the ordinary.

Spiritual experiences, though often subjective, can be profoundly transformative. They can provide a sense of peace, purpose, and interconnectedness that goes beyond the material world. These experiences are not easily explained by science, but they are nonetheless real and meaningful to those who have them.

The value of spirituality lies in its ability to address the existential questions that science alone cannot fully answer. It provides a framework for understanding the mysteries of life and death, the nature of the soul, and the potential for existence beyond the physical body. In this way, spirituality complements science, offering insights into the aspects of reality that lie beyond the empirical.

The Convergence of Science and Spirituality

In recent years, there has been a growing recognition that science and spirituality need not be mutually exclusive. Instead, they can be seen as complementary approaches to understanding the universe—two sides of the same coin. This convergence is reflected in the work of scientists and thinkers who seek to integrate scientific knowledge with spiritual wisdom, exploring the common ground between these two domains.

In a world increasingly driven by scientific discovery, spirituality offers a way to find meaning and purpose beyond the material (Smith, 2009). At the same time, scientific concepts, particularly in physics and cosmology, have increasingly influenced spiritual beliefs, leading to new interpretations of ancient religious ideas (Capra, 1975). This intersection of science and spirituality invites us to explore the mysteries of the universe with an open mind, considering the possibilities that lie beyond the limits of current knowledge.

One area where science and spirituality intersect is in the study of consciousness. As mentioned earlier, consciousness remains one of the great mysteries of science, but it is also a central concern of spirituality. By combining scientific research with spiritual practices, we may gain new insights into the nature of consciousness and its relationship to the universe.

For example, the study of meditation and mindfulness—a practice rooted in spiritual traditions—has shown that these practices can have measurable effects on the brain and body, promoting mental and physical well-being. This suggests that spiritual practices can be understood, at least in part, through the lens of science, offering a bridge between the two domains.

Another area of convergence is the exploration of altered states of consciousness, such as those induced by DMT or experienced during near-death experiences. These states often provide insights that challenge our conventional understanding of reality, suggesting that there may be more to existence than what can be observed in the physical world. By studying these experiences scientifically, we may uncover new dimensions of reality that align with spiritual teachings about the nature of the soul, the afterlife, and the interconnectedness of all things.

The Search for Meaning in a Complex Universe

Ultimately, the convergence of science and spirituality is about the search for meaning in a complex and often mysterious universe. Both approaches offer valuable insights, but neither can fully explain the totality of existence on its own. By integrating the strengths of both science and spirituality, we can develop a more comprehensive understanding of the universe—one that honors both the material and the spiritual, the known and the unknown.

This search for meaning is not just an intellectual exercise; it is a deeply personal and transformative journey. It invites us to explore the mysteries of existence with curiosity, humility, and an open mind, recognizing that the answers we seek may lie beyond the boundaries of any single discipline.

As we continue to explore the intersections between science and spirituality, we may discover new ways of understanding ourselves, our place in the universe, and the nature of reality itself. This journey is ongoing, and it invites us to remain open to new possibilities, to question our assumptions, and to embrace the unknown with wonder and awe.

Science and Spirituality in Practice: Bridging the Gap

As we delve deeper into the convergence of science and spirituality, it's essential to consider how these two approaches can be integrated in practice. While they often operate in different spheres, there are ways in which scientific inquiry and spiritual exploration can inform and enhance one another, leading to a more holistic understanding of reality.

The Role of Meditation and Mindfulness

One of the most accessible and well-studied intersections of science and spirituality is the practice of meditation and mindfulness.

Originating from ancient spiritual traditions, these practices have been embraced by modern science for their profound effects on the brain and body. Research has shown that regular meditation can lead to structural changes in the brain, including increased gray matter density in areas associated with memory, learning, and emotional regulation. Mindfulness practices have been shown to reduce stress, improve focus, and enhance overall well-being.

What makes meditation and mindfulness particularly interesting from a scientific perspective is their ability to bridge the gap between subjective experience and objective measurement. While these practices are deeply rooted in spiritual traditions, they can be studied scientifically through neuroimaging, physiological monitoring, and psychological assessments. This dual approach allows for a richer understanding of how these practices influence both the mind and body, offering insights into the nature of consciousness and the potential for spiritual growth.

For many practitioners, meditation and mindfulness serve as a gateway to spiritual experiences—moments of deep connection, insight, and transcendence. These experiences, while subjective, are often described in terms that resonate with spiritual teachings: a sense of oneness with the universe, the dissolution of the ego, and a profound understanding of the interconnectedness of all things. By studying these experiences scientifically, we can begin to map the terrain of the spiritual mind, offering a framework for understanding how spiritual practices can lead to lasting psychological and physiological changes.

Exploring Altered States of Consciousness

Another area where science and spirituality intersect is in the study of altered states of consciousness. Whether induced by substances like DMT or experienced spontaneously during near-death experiences, these states offer a unique window into the nature of consciousness and the potential for accessing realities beyond our everyday perception.

As discussed in the previous chapter, DMT is often associated with experiences of otherworldly realms, encounters with non-human entities, and a profound sense of interconnectedness. These experiences challenge our understanding of reality, suggesting that consciousness may be capable of accessing dimensions or planes of existence that are normally hidden from view.

From a scientific perspective, altered states of consciousness provide an opportunity to study the brain in ways that are not possible during ordinary waking states. By examining the neural correlates of these experiences, researchers can gain insights into the mechanisms that underlie perception, cognition, and the sense of self. This research has the potential to uncover new aspects of brain function and to expand our understanding of how consciousness operates at both the individual and collective levels.

From a spiritual perspective, altered states of consciousness are often seen as gateways to higher knowledge, deeper understanding, and spiritual awakening. Many spiritual traditions, from shamanism to mysticism, have long recognized the value of these states for exploring the nature of the soul, the universe, and the divine. By integrating scientific research with spiritual practice, we can begin to explore the full spectrum of consciousness, from the mundane to the transcendent.

The Holographic Universe: A Unifying Theory?

One of the most intriguing ideas to emerge from the intersection of science and spirituality is the concept of the holographic universe. This theory, which has its roots in both quantum mechanics and information theory, suggests that the entire universe can be seen as a hologram—a two-dimensional surface that encodes all the information needed to describe a three-dimensional reality.

The holographic principle, as it is known in physics, was initially proposed to resolve the paradoxes associated with black holes and information loss. It posits that the information contained within a black hole's event horizon is enough to describe everything that happens inside the black hole, effectively encoding the three-dimensional interior on a two-dimensional surface.

What makes the holographic principle particularly compelling is its potential to bridge the gap between science and spirituality. If the universe is indeed a hologram, this suggests that our perception of reality is fundamentally shaped by the way information is encoded and processed. In this view, the material world is not the ultimate reality, but rather a projection of a deeper, more fundamental layer of existence.

This idea resonates with many spiritual teachings, which often describe the physical world as an illusion or a veil that obscures the true nature of reality. In this sense, the holographic universe theory provides a scientific framework for understanding concepts that have long been central to spiritual traditions, such as the illusion of separateness, the interconnectedness of all things, and the existence of higher dimensions or planes of reality.

By exploring the holographic principle and its implications, we may find new ways to integrate scientific and spiritual knowledge, offering a more unified and comprehensive understanding of the universe and our place within it.

The Ethics of Scientific and Spiritual Integration

As science and spirituality continue to converge, it is crucial to consider the ethical implications of this integration. Both science and spirituality have the power to shape our understanding of the world and to influence the decisions we make as individuals and as a society. As such, the integration of these two approaches must be guided by a strong ethical framework that prioritizes the well-being of all beings and the planet as a whole.

One of the key ethical considerations is the potential for misuse or exploitation of scientific and spiritual knowledge. Just as scientific advancements can be used for both beneficial and harmful purposes, so too can spiritual practices and insights be co-opted for personal gain or manipulation. It is essential to ensure that the integration of science and spirituality is guided by principles of compassion, empathy, and respect for the autonomy and dignity of all individuals.

Another important consideration is the need for inclusivity and diversity in the integration of science and spirituality. Both fields have historically been dominated by certain cultural and intellectual traditions, often to the exclusion of others. As we seek to bring these two approaches together, it is important to recognize and value the contributions of all cultures, traditions, and perspectives. This includes honoring the wisdom of indigenous and non-Western spiritual practices, as well as embracing the diversity of scientific thought.

Finally, the integration of science and spirituality must be approached with humility and a recognition of the limits of human understanding. Both science and spirituality are ongoing, evolving processes, and neither can claim to have all the answers. By remaining open to new ideas and perspectives, and by approaching the mysteries of the universe with a sense of wonder and curiosity, we can continue to deepen our understanding of both the material and the spiritual dimensions of existence.

The Future of Science and Spirituality: A Unified Vision

As we look to the future, the convergence of science and spirituality holds the potential to transform our understanding of reality and to guide humanity toward a more holistic and harmonious way of living. This unified vision recognizes that the material and spiritual aspects of existence are not separate or opposed, but are instead different facets of the same underlying truth. By embracing both approaches, we can work toward a more integrated and meaningful understanding of the universe, one that honors both the physical and metaphysical dimensions of life.

The Role of Education in Bridging Science and Spirituality

One of the most important avenues for fostering the integration of science and spirituality is through education. By incorporating both scientific and spiritual perspectives into educational curricula, we can encourage students to explore the full range of human knowledge and experience. This approach can help to break down the false dichotomy between science and spirituality, allowing young people to see how these two approaches can complement and enrich one another.

In practice, this might involve teaching scientific concepts alongside discussions of their philosophical and ethical implications, or exploring spiritual traditions in the context of their historical and cultural significance. It could also include hands-on experiences with practices such as meditation, mindfulness, and contemplation, alongside scientific experiments and critical thinking exercises. The goal is to create a learning environment that values both empirical inquiry and personal reflection, fostering a well-rounded and open-minded approach to understanding the world.

By educating future generations in this way, we can help to cultivate a new kind of thinker—one who is comfortable navigating both the material and spiritual realms, and who is capable of integrating these perspectives to solve complex problems and contribute to the betterment of society.

Scientific and Spiritual Collaboration

Another promising development in the future of science and spirituality is the increasing collaboration between scientists and spiritual practitioners. As these two communities come together, they can share insights and methods that can lead to new discoveries and a deeper understanding of the universe.

For example, neuroscientists studying the effects of meditation on the brain have benefited greatly from the collaboration with experienced meditators and spiritual teachers. By working together, they have been able to design more effective experiments, interpret results in a more nuanced way, and generate new hypotheses about the nature of consciousness and the mind.

Similarly, spiritual practitioners can benefit from the insights and tools provided by science. Understanding the physiological and neurological effects of spiritual practices can help practitioners refine their techniques and achieve more consistent results. Additionally, the application of scientific methods to spiritual questions can lend credibility to these practices, helping to bridge the gap between spiritual and secular communities.

These collaborations are likely to expand in the coming years, as more researchers and practitioners recognize the value of integrating their knowledge and expertise. The result could be a new era of discovery and understanding, where science and spirituality work hand in hand to explore the deepest mysteries of existence.

A Vision for a Unified Future

The future of science and spirituality is not about choosing one path over the other, but about finding ways to integrate these approaches into a unified vision of reality. This vision recognizes that science and spirituality are not separate domains, but are instead complementary ways of exploring the same fundamental truths.

In this unified future, science and spirituality work together to address the most pressing challenges facing humanity—from environmental degradation and social inequality to the quest for meaning and purpose in an increasingly complex world. By embracing both the empirical and the intuitive, the material and the spiritual, we can develop new solutions that are both grounded in evidence and infused with wisdom.

This vision also invites us to rethink our relationship with the universe. Rather than seeing ourselves as isolated individuals in a mechanistic world, we can begin to see ourselves as interconnected beings, part of a vast and dynamic web of life. This shift in perspective can lead to a greater sense of responsibility for our actions, a deeper appreciation for the beauty and complexity of the world, and a renewed commitment to living in harmony with all beings.

In conclusion, the intersection of science and spirituality offers a path forward for humanity—one that honors both our rational minds and our spiritual hearts. As we continue to explore this convergence, we may find that the answers to the greatest mysteries of existence lie not in choosing one path over the other, but in embracing the richness and diversity of both.

Chapter 7: Black Holes, Quantum Mechanics, and the Nature of Reality

Introduction to Black Holes and Quantum Mechanics

Black holes and quantum mechanics represent two of the most profound and mysterious areas of modern physics. Black holes, with their intense gravitational pull, challenge our understanding of space and time, while quantum mechanics reveals the strange and counterintuitive behavior of particles at the smallest scales. These two areas of study, though seemingly separate, are deeply interconnected, and their intersection may hold the key to unlocking the ultimate nature of reality (Hawking, 1974; Thorne, 1994).

This chapter will explore the relationship between black holes and quantum mechanics, delving into the concept of the information paradox, the potential for black holes to serve as portals to other dimensions, and the broader implications for our understanding of the universe. By examining these topics, we will consider how they align with the insights and visions explored in previous chapters, particularly the idea of information transmission through black holes and the "101" message of universal interconnectedness.

The Information Paradox and the Holographic Principle

One of the central mysteries of black holes is the information paradox. According to classical physics, any information about matter that falls into a black hole is lost forever, as nothing can escape the event horizon. However, this idea conflicts with the principles of quantum mechanics, which assert that information must be conserved. The apparent loss of information in a black hole challenges our understanding of how the universe operates at both the quantum and cosmic scales (Page, 1993; Preskill, 1992).

To address this paradox, physicists have proposed several theories, one of the most compelling being the holographic principle. This principle suggests that the information about everything that falls into a black hole is not lost but is instead encoded on the event horizon—the boundary of the black hole. In this view, the event horizon acts like a two-dimensional surface that stores information about the three-dimensional space inside the black hole, much like a hologram (Bousso, 2002).

The holographic principle has profound implications for our understanding of the universe. It suggests that our perception of three-dimensional space may be an illusion, a projection of information encoded on a two-dimensional surface. This idea aligns with the concept that the universe itself could be a hologram, where the reality we experience is a projection of a deeper, more fundamental layer of existence (Susskind, 2008).

If black holes do indeed store information in this way, it raises the possibility that they could also serve as conduits for transmitting information across time and space, as explored in Chapter 3. The holographic nature of black holes could mean that the information they contain is not bound by the traditional constraints of space and time, allowing it to be accessed or transmitted in ways that challenge our conventional understanding of reality (Hawking, 1974).

Quantum Entanglement and Black Holes

Quantum entanglement is another key concept in understanding the relationship between black holes and the fundamental nature of reality. Entanglement occurs when two or more particles become linked in such a way that the state of one particle is instantaneously connected to the state of the other, regardless of the distance between them. This phenomenon suggests that there is a deep interconnectedness at the quantum level, where the separation between particles is an illusion (Einstein et al., 1935).

In recent years, physicists have begun to explore the idea that black holes might be connected through quantum entanglement. This concept, known as the ER=EPR conjecture (named after physicists Einstein, Rosen, and Podolsky), proposes that entangled particles might be linked by microscopic wormholes, or Einstein-Rosen bridges, within black holes. These wormholes could allow information to travel between black holes or even across different points in space-time, potentially offering a solution to the information paradox (Maldacena & Susskind, 2013).

The idea that black holes could be connected through quantum entanglement opens up new possibilities for understanding the nature of space, time, and reality itself. If black holes are indeed linked in this way, it suggests that the universe is far more interconnected than we currently realize. This interconnectedness could extend beyond the physical realm, offering a potential explanation for the "101" message and the idea that all things in the universe are fundamentally connected (Hawking, 1974; Susskind, 2008).

Black Holes as Portals to Other Dimensions

Another intriguing possibility is that black holes could serve as portals to other dimensions or parallel universes. According to some interpretations of string theory and other advanced theories in physics, black holes might be gateways to higher-dimensional spaces or alternate realities. These theories suggest that what we perceive as a black hole's singularity—the point where space and time break down—could be a passage to another universe or a different layer of reality (Greene, 1999; Polchinski, 1998).

If black holes do indeed connect to other dimensions, it raises the possibility that they could be used to transmit information or even matter across different realms of existence. This idea aligns with the vision of black holes as conduits for information transmission, as discussed in previous chapters. The possibility of accessing other dimensions through black holes also resonates with the idea that our perception of reality is limited and that there are aspects of the universe that we cannot directly observe or comprehend (Thorne, 1994; Susskind, 2008).

The concept of black holes as portals to other dimensions challenges our understanding of the universe and invites us to consider the possibility that there is far more to reality than what we can perceive. It suggests that the boundaries between different layers of existence may be more fluid than we realize, and that black holes could be the key to exploring these hidden dimensions (Greene, 1999; Maldacena & Susskind, 2013).

The Role of Black Holes in the Evolution of the Universe

Black holes are not merely isolated anomalies in the fabric of the cosmos; they play a critical role in the evolution of the universe. Their immense gravitational pull shapes the structure of galaxies, influences the formation of stars, and even affects the distribution of matter on a cosmic scale. As we delve deeper into the nature of black holes, we begin to see them not just as destructive forces, but as essential components in the dynamic processes that govern the universe (Hawking, 1974).

Black Holes and Galaxy Formation

One of the most significant roles black holes play is in the formation and evolution of galaxies. Nearly every large galaxy, including our own Milky Way, is believed to harbor a supermassive black hole at its center. These black holes, with masses millions or even billions of times that of our sun, exert a powerful influence on their surroundings, shaping the orbits of stars and the distribution of interstellar gas (Kormendy & Ho, 2013).

The relationship between supermassive black holes and their host galaxies is a subject of intense study in modern astrophysics. Observations suggest a close correlation between the mass of a galaxy's central black hole and the properties of the galaxy itself, such as its bulge mass and velocity dispersion. This correlation, known as the M-sigma relation, implies that the growth of black holes and the evolution of galaxies are intimately linked (Ferrarese & Merritt, 2000).

One hypothesis is that the energy released by matter falling into a supermassive black hole—through a process known as accretion—can regulate the growth of the galaxy. As matter spirals into the black hole, it heats up and emits radiation, which can drive powerful outflows of gas and dust from the galaxy's core. These outflows, in turn, can suppress star formation by blowing away the gas needed to form new stars, effectively controlling the growth of the galaxy. This feedback mechanism, known as "quasar mode" feedback, suggests that black holes play a crucial role in the evolution of galaxies, shaping their structure and determining their fate (Silk & Rees, 1998).

Hawking Radiation and the Lifespan of Black Holes

While black holes are often thought of as eternal, they are not, according to Stephen Hawking's groundbreaking theory. Hawking proposed that black holes can slowly lose mass and energy over time through a process now known as Hawking radiation. This radiation arises from quantum effects near the event horizon, where particle-antiparticle pairs spontaneously form and one of the particles escapes into space, while the other falls into the black hole, effectively reducing its mass (Hawking, 1974).

Hawking radiation suggests that black holes are not permanent fixtures in the universe; instead, they have finite lifespans. For a stellar-mass black hole, this evaporation process would take far longer than the current age of the universe, but for much smaller black holes, the process could happen much more quickly. The eventual evaporation of black holes raises profound questions about the ultimate fate of the universe and the role that black holes play in its long-term evolution (Hawking, 1974).

The concept of Hawking radiation also ties into the information paradox discussed earlier. If black holes can evaporate completely, what happens to the information that was stored within them? Does it get lost forever, or is it somehow preserved in the radiation emitted as the black hole evaporates? These questions remain at the forefront of theoretical physics and challenge our understanding of the fundamental principles that govern the universe (Page, 1993; Preskill, 1992).

Black Holes and the Arrow of Time

The study of black holes also intersects with one of the most profound questions in physics: the nature of time. In classical mechanics, time is considered a linear, unidirectional flow from the past to the future. However, the laws of physics at the quantum level are generally time-symmetric, meaning they do not inherently distinguish between past and future. This raises the question of why time seems to move in only one direction in our everyday experience—a concept known as the "arrow of time" (Price, 1996).

Black holes, with their extreme gravitational fields, offer a unique laboratory for exploring the nature of time. The intense warping of spacetime near a black hole's event horizon can cause time to appear to slow down relative to an outside observer—a phenomenon predicted by Einstein's theory of general relativity. This time dilation effect becomes so pronounced near the event horizon that, to an outside observer, time seems to almost stop (Thorne, 1994).

The concept of black holes influencing the flow of time leads to intriguing possibilities, particularly when considering the idea of information transmission through black holes. If black holes can affect the passage of time, could they also manipulate or transmit information across different points in time, as speculated in earlier chapters? This idea challenges our conventional understanding of causality and invites us to reconsider the nature of time itself (Susskind, 2008).

Moreover, the relationship between black holes and the arrow of time may offer insights into the ultimate fate of the universe. Some theories suggest that as the universe evolves, black holes will play a crucial role in determining its final state, potentially leading to a universe where time itself ceases to exist in the way we currently understand it. This concept ties into the idea of entropy—the measure of disorder in a system—and the eventual "heat death" of the universe, where all energy is evenly distributed, and no further work can be done (Penrose, 1989).

The Philosophical Implications of Black Holes

The study of black holes is not just a scientific endeavor; it also raises profound philosophical questions about the nature of existence, reality, and the universe itself. Black holes challenge our understanding of fundamental concepts such as space, time, and causality, leading us to reconsider what we know about the universe and our place within it (Thorne, 1994).

For instance, the idea that black holes could serve as gateways to other dimensions or parallel universes invites us to think about the nature of reality in new ways. If our universe is just one of many, and black holes are the portals that connect them, what does that say about the nature of existence? Are we living in a multiverse, where countless other realities exist alongside our own? And if so, what is the purpose of our universe within this larger, cosmic framework (Greene, 1999)?

The concept of the information paradox also touches on deeper philosophical questions about the nature of knowledge and the limits of human understanding. If information can be lost in a black hole, what does that say about the possibility of ever achieving a complete understanding of the universe? Are there aspects of reality that are fundamentally unknowable, hidden forever behind the event horizon of a black hole (Hawking, 1974; Preskill, 1992)?

Finally, the study of black holes invites us to reflect on the nature of existence itself. If black holes can evaporate through Hawking radiation, eventually disappearing from the universe, what does that say about the impermanence of all things? Does the eventual dissolution of black holes suggest that everything in the universe, no matter how massive or powerful, is ultimately transient? And if so, what does that mean for our understanding of life, death, and the nature of being (Hawking, 1974; Page, 1993)?

The Limits of Human Knowledge and the Mystery of Black Holes

As we continue to explore the enigmatic nature of black holes, it becomes increasingly clear that they represent one of the greatest challenges to our understanding of the universe. Despite the remarkable progress made in astrophysics and quantum mechanics, black holes remain shrouded in mystery, their true nature eluding even the most advanced theories. This chapter concludes by reflecting on the limits of human knowledge and the profound mysteries that black holes embody.

The Event Horizon: The Boundary of Knowledge

The event horizon of a black hole is often described as the "point of no return." It marks the boundary beyond which nothing, not even light, can escape the gravitational pull of the black hole. In a sense, the event horizon represents the ultimate limit of our knowledge. Once information crosses this boundary, it is effectively lost to the outside universe, rendering the interior of the black hole inaccessible to observation (Hawking, 1974).

This inaccessibility raises fundamental questions about the nature of reality and our ability to understand it. If we cannot observe or measure what happens inside a black hole, how can we claim to have a complete understanding of the universe? The event horizon challenges the very foundations of empirical science, which relies on observation and measurement to build theories and models. It suggests that there may be aspects of reality that are fundamentally unknowable, hidden forever from human inquiry (Thorne, 1994).

The concept of the event horizon also invites us to consider the limitations of human perception. Just as the event horizon marks the boundary of a black hole, so too does it symbolize the boundaries of our own knowledge and understanding. Beyond this boundary lies the unknown, the realm of possibilities that we can only speculate about but never fully comprehend (Hawking, 1974; Price, 1996).

The Singularity: The Edge of Understanding

At the center of a black hole lies the singularity—a point of infinite density where the laws of physics as we know them break down. The singularity represents the ultimate challenge to our understanding of the universe. It is a place where space and time cease to have meaning, where the very fabric of reality is warped beyond recognition (Penrose, 1989).

The existence of singularities suggests that our current theories, such as general relativity, are incomplete. While general relativity successfully describes the behavior of objects in strong gravitational fields, it fails to account for the quantum effects that become significant at the scale of the singularity. This discrepancy points to the need for a unified theory of quantum gravity—one that can reconcile the principles of quantum mechanics with the curvature of spacetime described by general relativity (Greene, 1999; Thorne, 1994).

The search for a theory of quantum gravity is one of the most pressing challenges in modern physics. Such a theory could provide the key to understanding not only black holes but also the origin of the universe itself. However, despite decades of research, a complete theory of quantum gravity remains elusive, highlighting the limits of our current knowledge (Penrose, 1989).

The singularity at the heart of a black hole is a reminder that there are still many unanswered questions about the nature of reality. It serves as a symbol of the mysteries that lie at the edge of our understanding, challenging us to push the boundaries of science and explore new frontiers of knowledge (Hawking, 1974; Thorne, 1994).

The Multiverse Hypothesis and Black Holes

One of the most speculative and intriguing ideas related to black holes is the multiverse hypothesis—the idea that our universe is just one of many, each existing within its own distinct realm of reality. According to some interpretations of string theory and other advanced theories in physics, black holes could serve as gateways to these alternate universes, connecting different regions of the multiverse through their singularities (Greene, 1999; Polchinski, 1998).

The multiverse hypothesis suggests that black holes might not be the dead ends we once thought them to be. Instead, they could be portals to other dimensions or parallel realities, where the laws of physics may be different from those in our own universe. This idea challenges our understanding of space and time, suggesting that the universe we observe is just one part of a much larger and more complex reality (Susskind, 2008).

The implications of the multiverse hypothesis are profound. If true, it would mean that our universe is not unique, but rather one of countless other universes, each with its own set of physical laws and constants. It would also imply that the information lost in a black hole in our universe might reappear in another, potentially offering a solution to the information paradox (Greene, 1999; Hawking, 1974).

However, the multiverse hypothesis remains speculative, and there is currently no empirical evidence to support it. Nonetheless, it represents one of the most exciting and imaginative frontiers in modern physics, pushing the boundaries of our understanding and challenging us to think about the nature of reality in new and profound ways (Thorne, 1994; Polchinski, 1998).

The Philosophical Reflection on Black Holes and Knowledge

The study of black holes forces us to confront the limits of human knowledge and the mysteries that lie beyond our understanding. It challenges us to think about the nature of reality, the boundaries of our perception, and the possibility that there are aspects of the universe that we may never fully comprehend (Hawking, 1974).

This reflection on the limits of knowledge invites us to adopt a stance of humility in the face of the unknown. As we explore the mysteries of the universe, we must acknowledge that there may be questions that science cannot answer, phenomena that defy explanation, and realities that lie beyond the reach of our theories and models (Price, 1996).

At the same time, the mysteries of black holes also inspire us to continue our quest for knowledge, to push the boundaries of science, and to explore new frontiers of understanding. They remind us that the universe is a vast and complex place, filled with wonders that challenge our imagination and invite us to expand our horizons (Penrose, 1989).

In the end, the study of black holes is not just about understanding the physical universe; it is also about exploring the deeper questions of existence, reality, and the nature of knowledge itself. As we continue to probe the mysteries of black holes, we are also exploring the mysteries of our own consciousness, our place in the cosmos, and the limits of what it means to know (Thorne, 1994).

Chapter 8: Reflections on Mental Illness and the Search for Meaning

Introduction to Mental Illness and the Journey of Understanding

Mental illness, particularly conditions like bipolar disorder, can profoundly shape one's perception of reality and the search for meaning in life. For those who experience the intense highs and lows associated with bipolar disorder, the boundaries between the ordinary and the extraordinary, the real and the imagined, can become blurred. This chapter explores the intersection of mental illness, the quest for understanding, and the insights that can emerge from altered states of consciousness.

Throughout this narrative, my experiences with bipolar disorder have played a central role in shaping my thoughts, visions, and understanding of the universe. These experiences have led to profound insights, but they have also brought challenges and uncertainties. In this chapter, I reflect on the nature of mental illness, the lessons it has taught me, and how it has influenced my search for meaning and truth.

The Nature of Bipolar Disorder

Bipolar disorder is a mental health condition characterized by extreme mood swings, including manic episodes of elevated mood and energy and depressive episodes of deep sadness and lethargy. These mood swings can be intense and disruptive, affecting every aspect of a person's life, from their relationships and work to their self-perception and worldview (American Psychiatric Association, 2013).

During manic episodes, individuals with bipolar disorder may experience heightened creativity, energy, and a sense of connection with the universe. Thoughts race, ideas flow, and the mind seems to operate at an accelerated pace. These periods can lead to brilliant insights and creative breakthroughs, but they can also result in impulsive decisions, risky behavior, and a distorted sense of reality (Jamison, 1993).

On the other hand, depressive episodes can bring feelings of hopelessness, worthlessness, and profound despair. The energy and creativity of the manic phase are replaced by a heavy darkness, making it difficult to engage with the world or find meaning in anything. These swings between extremes can be exhausting and disorienting, making it challenging to maintain a stable sense of self and reality (American Psychiatric Association, 2013).

For many people with bipolar disorder, these mood swings are not just emotional or psychological states; they are also experiences that shape their understanding of the world and their place within it. The heightened states of mania, in particular, can lead to experiences that feel profound, spiritual, or revelatory, even if they are later questioned or reconsidered in the light of a more stable mood.

The "Premonition" and Its Impact

One of the most striking experiences I had during a manic episode was what I perceived as a "premonition" about my termination from a previous job. During an intense manic state, I felt as though I was being communicated with by an intelligence beyond human understanding, conveying ideas and concepts in a way that transcended ordinary conversation (Bentall, 2003).

This premonition included specific details about my future, including a confrontation with a superior, my subsequent termination, and the next job I would hold. Remarkably, these events unfolded almost exactly as predicted, leading me to question the nature of the experience. Was it merely a product of my manic state, a coincidence, or something more—a glimpse into a deeper reality? (Bentall, 2003).

The experience of the premonition has had a lasting impact on my perception of reality and my search for meaning. On one hand, I recognize the possibility that the experience was shaped by my mental illness, and that the accuracy of the predictions may have been a coincidence or the result of subconscious pattern recognition (Bentall, 2003). On the other hand, the experience felt undeniably real at the time, and its accuracy has led me to consider the possibility that there may be more to reality than we typically perceive.

This duality—between the recognition of mental illness and the sense of connection to something greater—has been a central theme in my journey. It has led me to explore the boundaries of what we can know and understand, and to consider the possibility that altered states of consciousness, whether induced by mental illness or other means, may offer glimpses into aspects of reality that are ordinarily hidden from view.

Mental Illness and the Search for Truth

There is a long history of association between mental illness and creativity, with many artists, writers, and thinkers attributing their most profound work to their experiences with conditions like bipolar disorder. The intense emotions, unique perspectives, and altered states of consciousness that accompany mental illness can serve as powerful catalysts for creativity, leading to the creation of works that challenge conventional thinking and expand the boundaries of human understanding (Jamison, 1993).

The experience of mental illness, particularly conditions like bipolar disorder, raises profound questions about the nature of truth and reality. When the mind is in an altered state, whether through mania, depression, or other forms of mental illness, the boundary between what is real and what is imagined can become fluid. This fluidity challenges our conventional understanding of truth, inviting us to consider the possibility that reality is not as fixed or objective as we might assume.

For many people with bipolar disorder, the manic phase can bring experiences that feel deeply meaningful, even if they later seem irrational or delusional. These experiences often involve a heightened sense of connection with the universe, a feeling of being "in tune" with a greater reality, or the perception of receiving insights or messages from beyond. While these experiences can be disorienting and even dangerous, they can also be profoundly transformative, leading to new ways of thinking and understanding.

At the same time, the depressive phase can bring a sense of disconnection, isolation, and meaninglessness. The insights gained during mania may be questioned or dismissed, and the search for truth may feel like an impossible task. This cycle of highs and lows, of connection and disconnection, can make it difficult to find a stable sense of reality or to trust one's own perceptions.

In my own life, the creative impulses that arise during manic episodes have often led to periods of intense productivity and insight. These moments of heightened creativity have allowed me to explore new ideas, develop complex theories, and connect seemingly disparate concepts in ways that would not have been possible in a more stable state (Jamison, 1993). While these creative bursts can sometimes lead to unfinished projects or ideas that later seem impractical, they have also contributed to my growth as a thinker and a seeker of knowledge.

The Role of Skepticism and Open-Mindedness

In navigating the complexities of mental illness and the search for meaning, I have found it essential to balance skepticism with open-mindedness. On one hand, it is important to recognize the ways in which mental illness can distort perception and to approach extraordinary experiences with caution. On the other hand, it is also important to remain open to the possibility that these experiences, even if shaped by mental illness, may offer valuable insights into the nature of reality.

Skepticism serves as a grounding force, helping to keep the mind anchored in the known and the observable. It encourages critical thinking, rational analysis, and a careful consideration of evidence. In the context of mental illness, skepticism can help prevent the mind from becoming lost in delusion or fantasy, offering a way to evaluate experiences and discern what is likely to be true.

At the same time, open-mindedness allows for the exploration of new ideas, perspectives, and possibilities. It invites curiosity, creativity, and the willingness to question established beliefs. For those who experience altered states of consciousness, open-mindedness can provide a space to explore the insights gained during these states, even if they challenge conventional understanding.

In my own journey, I have found that maintaining this balance between skepticism and open-mindedness has been crucial. It has allowed me to explore the profound experiences that have emerged from my manic episodes while also recognizing the limitations of my perceptions and the influence of my mental illness. This balance has helped me to navigate the complexities of my condition and to continue my search for truth and meaning in a way that is both grounded and expansive.

The Intersection of Mental Illness and Spiritual Experiences

The experiences of those with mental illness, particularly conditions like bipolar disorder, often blur the lines between what is considered spiritual and what is seen as symptomatic. This intersection raises profound questions about the nature of spiritual experiences and whether they might sometimes be enhanced or even triggered by mental states that alter perception.

Mania and Mysticism: A Fine Line

During manic episodes, individuals with bipolar disorder may experience what can only be described as mystical or spiritual states. These states often include a heightened sense of awareness, feelings of oneness with the universe, and a deep conviction that they have accessed hidden truths or insights. In many cases, these experiences are accompanied by intense emotions, such as euphoria, awe, or a sense of being chosen for a special purpose.

The intensity and nature of these experiences often mirror those described by mystics and spiritual seekers throughout history. Many religious traditions have accounts of individuals who, during periods of intense prayer, fasting, or meditation, enter states of consciousness that allow them to perceive realities beyond the ordinary. These experiences are often described as encounters with the divine, moments of enlightenment, or revelations of profound truths.

The similarities between manic states and mystical experiences raise intriguing questions about the nature of spirituality and its relationship to mental illness. Is it possible that mania can open the mind to experiences that are genuinely spiritual, or are these experiences simply the product of a mind in overdrive? Could the brain, under the influence of mania, be more attuned to aspects of reality that are usually hidden from view, or are these experiences merely delusions?

While these questions are difficult to answer definitively, they invite us to reconsider the boundaries between mental illness and spirituality. They suggest that the experiences of those with bipolar disorder, while shaped by their condition, may not be entirely disconnected from the broader human quest for understanding and connection with the universe.

The Search for Meaning Amidst Mental Illness

One of the most challenging aspects of living with mental illness is the search for meaning in experiences that can often feel disorienting, chaotic, or even terrifying. For those with bipolar disorder, the rapid shifts between mania and depression can make it difficult to find a stable sense of self or to understand the purpose of their experiences.

Yet, despite these challenges, many individuals with mental illness find that their condition leads them to ask deep and meaningful questions about life, existence, and the universe. The very intensity of their experiences can push them to explore the boundaries of what is known and to seek answers to the fundamental questions that have occupied humanity for millennia.

In my own experience, the heightened states of mania have often brought insights that feel profound, even if they are later questioned or reconsidered. These experiences have led me to explore concepts like the nature of reality, the interconnectedness of all things, and the possibility of life beyond the physical world. While I recognize that these insights are shaped by my mental illness, I also believe that they hold value and that they have contributed to my ongoing search for meaning.

The depressive episodes, while more difficult to navigate, have also played a role in this search. They have forced me to confront the darker aspects of existence—suffering, loss, and the feeling of being disconnected from the world. Yet, in grappling with these challenges, I have found a deeper understanding of the human condition and a greater empathy for others who are struggling.

Ultimately, the search for meaning amidst mental illness is a deeply personal journey, one that requires both courage and humility. It involves navigating the complexities of the mind, acknowledging the limitations of perception, and remaining open to the possibility that even the most difficult experiences can offer valuable lessons and insights.

Mental Illness, Creativity, and the Expansion of Consciousness

There is a long history of association between mental illness and creativity, with many artists, writers, and thinkers attributing their most profound work to their experiences with conditions like bipolar disorder. The intense emotions, unique perspectives, and altered states of consciousness that accompany mental illness can serve as powerful catalysts for creativity, leading to the creation of works that challenge conventional thinking and expand the boundaries of human understanding.

In my own life, the creative impulses that arise during manic episodes have often led to periods of intense productivity and insight. These moments of heightened creativity have allowed me to explore new ideas, develop complex theories, and connect seemingly disparate concepts in ways that would not have been possible in a more stable state. While these creative bursts can sometimes lead to unfinished projects or ideas that later seem impractical, they have also contributed to my growth as a thinker and a seeker of knowledge.

The connection between mental illness and creativity suggests that altered states of consciousness, whether induced by mental illness, meditation, or other means, can offer unique perspectives on the world. These perspectives, while sometimes difficult to integrate into everyday life, can lead to breakthroughs in understanding, artistic expression, and the expansion of consciousness.

However, it is also important to recognize the challenges that come with this connection. The creative energy that accompanies mental illness can be difficult to harness and sustain, and it can sometimes lead to burnout, frustration, or feelings of inadequacy. Moreover, the same insights that feel profound and transformative in a manic state can be dismissed or forgotten during periods of depression, leading to a sense of disconnection from one's own creative potential.

Despite these challenges, I believe that the creative impulses associated with mental illness can be a source of strength and inspiration. They offer a way to channel the intense emotions and experiences of mental illness into something meaningful, whether through art, writing, or the development of new ideas. In this way, creativity becomes not just a coping mechanism, but a path to greater understanding and a means of contributing to the broader human quest for knowledge and meaning.

The Role of Science in Understanding Mental Illness

While spirituality and personal experience offer valuable perspectives on mental illness, science provides essential tools for understanding the biological, psychological, and social aspects of these conditions. Advances in neuroscience, psychology, and psychiatry have shed light on the complex mechanisms that underlie mental illnesses like bipolar disorder, offering insights that can inform treatment, reduce stigma, and improve quality of life.

The Neuroscience of Bipolar Disorder

Bipolar disorder is a complex condition that affects mood regulation, thought processes, and behavior. While the exact causes of bipolar disorder are not fully understood, research suggests that it involves a combination of genetic, environmental, and neurobiological factors.

One of the key areas of research in understanding bipolar disorder is the study of brain structure and function. Neuroimaging studies have revealed differences in the brain regions involved in emotion regulation, such as the amygdala, prefrontal cortex, and hippocampus. These differences may contribute to the mood swings and emotional instability characteristic of bipolar disorder.

In addition to structural differences, bipolar disorder is associated with dysregulation of neurotransmitters—chemicals in the brain that transmit signals between neurons. Dopamine, serotonin, and norepinephrine are among the neurotransmitters implicated in the mood swings of bipolar disorder. For example, elevated levels of dopamine may contribute to the heightened energy and euphoria experienced during manic episodes, while reduced serotonin levels may be linked to the depressive phases.

Understanding these neurobiological factors is crucial for developing effective treatments for bipolar disorder. Medications such as mood stabilizers, antipsychotics, and antidepressants target specific neurotransmitter systems, helping to regulate mood and reduce the severity of symptoms. Ongoing research in neuroscience aims to refine these treatments and explore new approaches, such as neuromodulation and personalized medicine, to better address the needs of individuals with bipolar disorder.

The Psychological and Social Dimensions of Mental Illness

While neuroscience provides insights into the biological underpinnings of mental illness, psychology offers a complementary perspective on the cognitive, emotional, and behavioral aspects of these conditions. Cognitive-behavioral therapy (CBT), for example, is a widely used approach that helps individuals with bipolar disorder identify and challenge negative thought patterns, develop coping strategies, and build resilience.

Cognitive-behavioral therapy (CBT), for example, is a widely used approach that helps individuals with bipolar disorder identify and challenge negative thought patterns, develop coping strategies, and build resilience. CBT focuses on the connection between thoughts, emotions, and behaviors, offering practical tools for managing mood swings and reducing the impact of bipolar disorder on daily life.

Another important psychological approach is mindfulness-based therapy, which encourages individuals to develop a non-judgmental awareness of their thoughts and emotions. Mindfulness practices can help individuals with bipolar disorder stay grounded in the present moment, reduce stress, and cultivate a greater sense of control over their emotional responses. These practices can be particularly valuable in managing the intense emotions that often accompany manic and depressive episodes.

In addition to individual therapy, social support is a critical factor in the management of bipolar disorder. Relationships with family, friends, and mental health professionals can provide essential emotional support, reduce feelings of isolation, and offer practical assistance in navigating the challenges of the condition. Group therapy and peer support groups can also be valuable, providing a sense of community and shared experience that can help individuals feel less alone in their struggles.

The social dimension of mental illness also includes the impact of stigma. Despite growing awareness and understanding of mental health issues, stigma remains a significant barrier to seeking help and receiving appropriate care. Stigma can lead to feelings of shame, self-doubt, and isolation, making it harder for individuals with bipolar disorder to reach out for support. Addressing stigma through education, advocacy, and open conversations about mental health is essential for creating a more supportive and understanding society.

Integrating Science and Personal Experience

For individuals living with bipolar disorder, integrating scientific knowledge with personal experience can be a powerful approach to understanding and managing the condition. While science offers valuable insights into the biological, psychological, and social aspects of bipolar disorder, personal experience provides a unique perspective on how these factors interact in the context of an individual's life.

For example, understanding the neurobiological basis of mood swings can help individuals make sense of their experiences and recognize that their symptoms are not a reflection of personal weakness or failure. Similarly, learning about effective psychological therapies can empower individuals to take an active role in their treatment, using evidence-based strategies to manage their symptoms and improve their quality of life.

At the same time, personal experience can inform and enrich scientific understanding of mental illness. By sharing their stories, individuals with bipolar disorder can help researchers, clinicians, and the broader public gain a deeper understanding of the lived experience of the condition. This can lead to more compassionate and effective approaches to treatment and support, as well as greater recognition of the strengths and insights that individuals with bipolar disorder can bring to the world.

The Ongoing Journey of Self-Discovery

Living with bipolar disorder is an ongoing journey of self-discovery, one that involves navigating the challenges of the condition while also exploring its potential for growth, creativity, and insight. For many individuals, this journey involves finding a balance between accepting the realities of the condition and seeking ways to transcend its limitations.

In my own experience, bipolar disorder has been both a source of struggle and a catalyst for exploration. The intense emotions, altered states of consciousness, and shifts in perspective that accompany the condition have led me to question the nature of reality, the meaning of life, and the boundaries of human understanding. While these questions are shaped by my mental illness, they are also part of a broader human quest for knowledge and meaning.

This journey has taught me the importance of self-compassion, the value of resilience, and the need for both skepticism and open-mindedness in the search for truth. It has also reinforced the idea that mental illness, while challenging, can be a source of strength and insight, offering unique perspectives on the world and the human condition.

As I continue to explore the intersection of science, spirituality, and personal experience, I remain committed to seeking a deeper understanding of bipolar disorder and its role in my life. This journey is not just about managing symptoms or finding answers; it is about embracing the complexity of the human experience and recognizing the potential for growth, connection, and meaning even in the face of adversity.

Conclusion of Chapter 8

Chapter 8 has explored the complex interplay between mental illness, spirituality, and the search for meaning. By examining the nature of bipolar disorder, the impact of manic and depressive episodes, and the role of science and personal experience in understanding the condition, this chapter has highlighted the challenges and opportunities that come with living with mental illness.

In the next chapter, we will return to the broader themes of consciousness, reality, and the universe, exploring how the insights gained from both science and spirituality can inform our understanding of existence and our place within it. This exploration will continue to weave together the threads of personal experience, scientific inquiry, and spiritual reflection, offering a holistic perspective on the nature of reality.

Chapter 9: Consciousness, Reality, and the Universe

Introduction to the Exploration of Consciousness and Reality

Consciousness and reality are two of the most profound and enigmatic subjects in the fields of science, philosophy, and spirituality. While we experience consciousness directly, understanding its nature—how it arises, what it is, and how it connects us to reality—remains one of the greatest challenges of human thought. Similarly, the nature of reality, including the fabric of space and time, the existence of multiple dimensions, and the possibility of parallel universes, continues to intrigue and puzzle scientists and philosophers alike.

In this chapter, we will delve into these interconnected topics, exploring how consciousness shapes our perception of reality and how the latest scientific discoveries challenge our understanding of the universe. We will examine the role of consciousness in the context of quantum mechanics, the implications of higher dimensions and parallel universes, and the potential for consciousness to extend beyond the physical body.

This exploration will continue to draw on the themes of science, spirituality, and personal experience, offering a holistic perspective on the mysteries of existence and our place within the cosmos.

Consciousness and Quantum Mechanics

Quantum mechanics, the branch of physics that deals with the behavior of particles at the smallest scales, has introduced a level of strangeness and uncertainty into our understanding of reality. Unlike classical physics, which describes a deterministic and predictable universe, quantum mechanics reveals a world where particles can exist in multiple states at once, where the act of observation can alter the outcome of an experiment, and where entanglement connects particles across vast distances instantaneously (Bohr, 1935; Heisenberg, 1958).

One of the most intriguing aspects of quantum mechanics is the role of the observer. The famous "double-slit experiment," for example, shows that particles such as electrons behave differently when observed than when unobserved. When unobserved, particles exhibit wave-like behavior, passing through both slits simultaneously. However, when observed, they behave like particles, passing through one slit or the other. This phenomenon, known as the "observer effect," suggests that consciousness itself may play a role in shaping reality (Bohr, 1935; Heisenberg, 1958).

The idea that consciousness could influence the physical world challenges the traditional scientific view of a universe that exists independently of the observer. It raises questions about the nature of reality and the relationship between mind and matter. Are we passive observers of a pre-existing reality, or do we play an active role in creating the world we experience? (Bohr, 1935; Heisenberg, 1958).

Some interpretations of quantum mechanics, such as the "many-worlds" interpretation, suggest that every possible outcome of a quantum event actually occurs in a separate, parallel universe. This implies that there may be countless parallel realities, each corresponding to a different set of choices and observations. If consciousness plays a role in determining which reality we experience, it suggests that our perception of reality is not fixed but is instead a dynamic and interactive process.

The Potential for Higher Dimensions and Parallel Universes

The concept of higher dimensions and parallel universes is not just a staple of science fiction—it is also a serious topic of scientific inquiry. String theory, one of the leading candidates for a unified theory of physics, posits the existence of multiple dimensions beyond the familiar three dimensions of space and one of time. According to string theory, the fundamental building blocks of the universe are not point-like particles but rather tiny, vibrating strings that exist in a higher-dimensional space (Kaku, 2000; Greene, 1999).

These extra dimensions, though hidden from our everyday experience, could have profound implications for our understanding of reality. They might explain the fundamental forces of nature, provide the basis for the unification of general relativity and quantum mechanics, and offer insights into the nature of black holes and the origin of the universe (Kaku, 2000; Greene, 1999).

The idea of parallel universes, or the multiverse, extends the concept of higher dimensions even further. In the multiverse hypothesis, our universe is just one of many, each with its own set of physical laws and constants. These parallel universes might exist in higher-dimensional spaces or could be separated from our own by vast distances in a cosmic landscape (Kaku, 2000; Greene, 1999).

The potential existence of parallel universes raises profound questions about the nature of reality and our place within it. If there are countless other universes, each with its own version of reality, what does that say about the uniqueness of our own universe? Could consciousness, in some way, traverse these different realities, accessing experiences or knowledge that lie beyond our ordinary perception? (Kaku, 2000; Greene, 1999).

The exploration of higher dimensions and parallel universes challenges our conventional understanding of space, time, and reality. It suggests that what we perceive as the "real" world may be just one slice of a much larger and more complex structure, and that our consciousness may be capable of interacting with these hidden dimensions in ways we do not yet fully understand.

Consciousness Beyond the Physical Body

Another area of inquiry that connects science, spirituality, and personal experience is the question of whether consciousness can exist independently of the physical body. This idea, while controversial, has been explored through various lines of research, including near-death experiences (NDEs), out-of-body experiences (OBEs), and the study of altered states of consciousness induced by substances like DMT (Moody, 1975; Greyson, 1983).

Near-death experiences, in particular, have been reported by individuals who have come close to death and then been revived. These experiences often include a sense of detachment from the physical body, encounters with deceased loved ones, journeys through tunnels of light, and a sense of merging with a universal consciousness (Moody, 1975; Greyson, 1983). While skeptics argue that these experiences can be explained by brain activity during the dying process, others suggest that they may provide evidence for the existence of consciousness beyond the physical body (Moody, 1975; Greyson, 1983).

Out-of-body experiences, where individuals report perceiving the world from a vantage point outside their physical body, also raise questions about the nature of consciousness. While OBEs can be induced through meditation, sensory deprivation, or certain substances, they are also reported spontaneously, often during moments of extreme stress or trauma. These experiences suggest that consciousness may not be entirely bound to the physical brain and body but may have the capacity to transcend physical limitations (Moody, 1975; Greyson, 1983).

The study of altered states of consciousness, whether through meditation, psychedelic substances, or other means, also offers insights into the potential for consciousness to access realities beyond the physical. As discussed in Chapter 4, substances like DMT can induce experiences that feel profoundly real and meaningful, leading to encounters with beings or entities that seem to exist in a different dimension or plane of reality (Strassman, 2001).

The possibility that consciousness can exist beyond the physical body challenges the materialist view of the mind as a product of brain activity. It opens up the possibility that consciousness is a fundamental aspect of the universe, one that is not limited by physical boundaries but is instead interconnected with the fabric of reality itself.

The Search for a Unified Understanding of Consciousness and Reality

The exploration of consciousness and reality invites us to seek a unified understanding that integrates insights from science, spirituality, and personal experience. While science provides powerful tools for understanding the physical world, spirituality offers a framework for exploring the deeper, non-material aspects of existence. Personal experience, particularly in altered states of consciousness, can provide direct insights into the nature of reality that may not be accessible through ordinary perception.

In this search for a unified understanding, it is important to remain open to the possibility that reality is far more complex and multifaceted than we currently understand. The intersection of quantum mechanics, higher dimensions, and consciousness suggests that the universe may be a far stranger and more mysterious place than we can imagine. It also suggests that consciousness itself may be a key to unlocking these mysteries, offering a bridge between the physical and the metaphysical, the known and the unknown.

As we continue to explore these questions, it is essential to approach them with a sense of curiosity, humility, and wonder. The search for understanding is not just an intellectual pursuit; it is also a deeply personal and transformative journey, one that invites us to question our assumptions, expand our horizons, and embrace the mystery of existence.

Conclusion of Chapter 9

Chapter 9 has explored the interconnectedness of consciousness, reality, and the universe, drawing on insights from quantum mechanics, the potential for higher dimensions and parallel universes, and the possibility of consciousness extending beyond the physical body. This exploration challenges our conventional understanding of reality and invites us to consider the profound mysteries that lie at the intersection of science, spirituality, and personal experience.

In the next chapter, we will continue to build on these themes, exploring the implications of these ideas for our understanding of existence, the nature of life and death, and the search for meaning in a complex and interconnected universe. This journey will continue to weave together the threads of scientific inquiry, spiritual reflection, and personal experience, offering a holistic perspective on the nature of reality.

Chapter 10: The Nature of Life, Death, and the Afterlife

Introduction to Life, Death, and the Afterlife

Life, death, and the afterlife are among the most profound and universal concerns of human existence. Across cultures and throughout history, these topics have been explored through religion, philosophy, and science, each offering different perspectives on what it means to live, die, and what—if anything—comes after death. This chapter will explore these themes in the context of the insights gained from earlier chapters, including the nature of consciousness, the potential for higher dimensions, and the mysteries of the universe.

We will examine various cultural and religious views on life and death, the scientific understanding of these processes, and the possibilities for what might exist beyond physical death. This exploration will also consider how the concepts of quantum mechanics, consciousness, and higher dimensions could inform our understanding of the afterlife, offering a bridge between scientific inquiry and spiritual belief.

Cultural and Religious Perspectives on Life and Death

Throughout history, different cultures and religions have offered a wide range of beliefs and practices surrounding life, death, and the afterlife. These beliefs are often deeply rooted in the spiritual and metaphysical views of the world, shaping how individuals and communities understand their existence and their relationship with the universe (Eliade, 1959).

In many religious traditions, the afterlife is seen as a continuation of the soul's journey beyond the physical world. For example, in Christianity, the afterlife is often conceptualized as a heaven or hell, where souls are rewarded or punished based on their actions in life (The Holy Bible). In Hinduism and Buddhism, the concept of reincarnation suggests that the soul undergoes a cycle of birth, death, and rebirth, with each lifetime offering an opportunity for spiritual growth and the eventual attainment of enlightenment (Bhagavad Gita; The Dhammapada).

Indigenous cultures also have rich traditions surrounding life and death, often viewing the afterlife as a return to a spiritual homeland or a continuation of life in a different form. These beliefs are often intertwined with a deep connection to the natural world and a sense of continuity between the physical and spiritual realms (Deloria, 1994).

While the specifics of these beliefs vary widely, they share a common theme: the idea that life and death are part of a larger, ongoing process that extends beyond the physical body. This view contrasts with the materialist perspective, which holds that consciousness is a product of brain activity and ceases with physical death (Dawkins, 2006). The materialist view, while dominant in modern science, leaves many questions unanswered, particularly regarding the nature of consciousness and the possibility of existence beyond the physical realm (Nagel, 2012).

The Scientific Understanding of Life and Death

From a scientific perspective, life and death are understood primarily in biological terms. Life is characterized by processes such as metabolism, growth, reproduction, and response to stimuli, all of which are governed by the biochemical machinery of cells (Mayr, 1982). Death, on the other hand, is defined by the cessation of these processes, leading to the breakdown of the body's systems and the eventual return of its components to the environment (Mayr, 1982).

The study of life and death from a scientific standpoint has led to remarkable advances in medicine, biology, and technology, allowing us to extend life, understand the causes of death, and explore the boundaries between the living and the non-living (Dawkins, 2006). Yet, despite these advances, science has struggled to explain certain phenomena that seem to transcend the biological understanding of life and death, such as near-death experiences, out-of-body experiences, and reports of consciousness continuing after clinical death (Greyson, 1983).

These phenomena challenge the materialist view of consciousness and raise questions about the nature of life and death that science has yet to fully address. For example, near-death experiences often involve vivid and detailed perceptions that occur while the brain is clinically inactive, suggesting that consciousness may not be entirely dependent on brain activity (Moody, 1975; Greyson, 1983). Similarly, reports of individuals perceiving events or details from outside their physical bodies during out-of-body experiences challenge our understanding of the relationship between consciousness and the physical body (Moody, 1975; Greyson, 1983).

While these experiences are often dismissed as hallucinations or the result of brain activity during trauma, they continue to inspire debate and research, particularly in the fields of neuroscience, psychology, and quantum physics. The possibility that consciousness could survive physical death, or that it could exist independently of the brain, remains an open question—one that touches on the deepest mysteries of existence (Greyson, 1983; Moody, 1975).

Quantum Mechanics and the Possibility of an Afterlife

As explored in previous chapters, quantum mechanics introduces a level of uncertainty and complexity into our understanding of reality that challenges the materialist view. Concepts such as quantum entanglement, the observer effect, and the potential existence of parallel universes suggest that the universe is far more interconnected and mysterious than we can fully comprehend (Heisenberg, 1958; Greene, 1999).

One of the intriguing possibilities raised by quantum mechanics is the idea that consciousness itself could be a fundamental aspect of the universe, not merely a byproduct of physical processes (Penrose & Hameroff, 1996). If consciousness is indeed fundamental, it might not be bound by the same constraints as the physical body, potentially allowing it to persist or even transition into other forms of existence after death (Penrose & Hameroff, 1996).

The idea that consciousness could continue after death aligns with certain interpretations of quantum mechanics, particularly those that involve the multiverse or higher dimensions. For example, if every possible outcome of a quantum event occurs in a separate parallel universe, it is conceivable that consciousness could "shift" into one of these alternate realities after death (Greene, 1999). This would imply that the death of the physical body in one universe does not necessarily mean the end of consciousness, but rather a transition to a different state of existence (Greene, 1999).

Another possibility is that consciousness could interact with or access higher dimensions, which are posited by theories such as string theory. If higher dimensions exist, they could provide a realm in which consciousness continues after physical death, existing in a form that is not bound by the limitations of the physical world (Kaku, 2000).

While these ideas remain speculative and are not widely accepted within the scientific community, they offer intriguing possibilities for understanding the afterlife from a perspective that integrates both science and spirituality. They suggest that the boundaries between life and death, and between the physical and non-physical realms, may be more fluid than we currently understand (Penrose & Hameroff, 1996; Greene, 1999).

The Search for Meaning in Life and Death

The exploration of life, death, and the afterlife is not just a scientific or philosophical endeavor; it is also a deeply personal and existential journey. The questions of what happens after we die, what it means to live a meaningful life, and how we should approach the inevitability of death are central to the human experience (Frankl, 1946).

For many people, the search for meaning in life and death is closely tied to their spiritual beliefs and practices. These beliefs provide a framework for understanding the purpose of life, the nature of the soul, and the possibility of an afterlife. They offer comfort in the face of mortality and help individuals find peace with the idea of death as a natural part of existence (Frankl, 1946; Eliade, 1959).

At the same time, the search for meaning can also be informed by scientific inquiry and personal experience. Understanding the biological processes of life and death, exploring the mysteries of consciousness, and reflecting on the experiences of others who have encountered death or near-death can provide valuable insights into the nature of existence and the possibilities that lie beyond (Moody, 1975; Greyson, 1983).

In my own journey, the search for meaning has been shaped by both my personal experiences and my exploration of the scientific and spiritual dimensions of life and death. The visions and insights gained during manic episodes, the contemplation of quantum mechanics and higher dimensions, and the reflections on my own mortality have all contributed to a deeper understanding of what it means to live and die (Jamison, 1993; Penrose & Hameroff, 1996).

Ultimately, the search for meaning in life and death is a journey that each individual must undertake for themselves. It involves grappling with the unknown, embracing the mysteries of existence, and finding a sense of purpose that resonates with one's own beliefs and experiences. Whether informed by science, spirituality, or a combination of both, this search is an essential part of what it means to be human.

Conclusion of Chapter 10

Chapter 10 has explored the nature of life, death, and the afterlife, drawing on insights from cultural and religious perspectives, scientific understanding, and the possibilities introduced by quantum mechanics and consciousness. This exploration highlights the profound and enduring questions that surround our existence and the mysteries that lie at the boundaries of life and death.

In the final chapter, we will synthesize the themes explored throughout this narrative, reflecting on the integration of science, spirituality, and personal experience, and considering how these insights can inform our understanding of the universe and our place within it. This conclusion will offer a holistic perspective on the nature of reality, the search for meaning, and the ongoing journey of exploration and discovery.

Chapter 11: Synthesis and Reflection on the Nature of Reality

Introduction to the Integration of Science, Spirituality, and Personal Experience

Throughout this narrative, we have explored a wide range of topics, from quantum mechanics and black holes to consciousness, mental illness, and the nature of life and death. These explorations have drawn on insights from science, spirituality, and personal experience, offering a holistic perspective on the mysteries of existence and our place within the universe (Capra, 1975; Wilber, 2000).

This chapter is not just a conclusion, but an invitation to continue the journey of exploration and discovery. The questions we have asked, the ideas we have considered, and the experiences we have reflected upon are all part of an ongoing quest to understand the universe and our role within it. As we bring these threads together, we will consider how they can guide us in our search for meaning, purpose, and a deeper connection with the cosmos (Capra, 1975; Wilber, 2000).

The Interconnectedness of All Things

One of the central themes that has emerged throughout this narrative is the idea of interconnectedness—the notion that all things in the universe are fundamentally linked, whether through the fabric of space-time, the principles of quantum mechanics, or the shared experiences of consciousness (Einstein, 1916; Zeilinger, 2010). This concept of interconnectedness is not just a scientific idea, but a spiritual one as well, resonating with the teachings of many religious and philosophical traditions (Capra, 1975; Wilber, 2000).

In quantum mechanics, the phenomenon of entanglement shows that particles can become instantaneously connected, no matter how far apart they are. This suggests that the universe is not composed of isolated, independent objects, but rather a web of interconnected relationships that transcend the limitations of space and time (Zeilinger, 2010). Similarly, the idea that consciousness itself may be a fundamental aspect of the universe implies that our minds are not separate from the world around us, but are deeply integrated with the fabric of reality (Capra, 1975).

Spiritual traditions, too, have long emphasized the interconnectedness of all things. Whether through the concept of a universal consciousness, the idea of the oneness of all life, or the belief in a divine force that permeates the cosmos, these teachings suggest that we are not isolated beings, but part of a larger, interconnected whole (Wilber, 2000). This perspective encourages us to see our lives as part of a greater narrative, one that connects us to each other, to nature, and to the universe itself (Capra, 1975).

Reflecting on this theme of interconnectedness invites us to consider how we can live in greater harmony with the world around us. It challenges us to think about the impact of our actions, the importance of compassion and empathy, and the need to cultivate a sense of unity and cooperation in our relationships, our communities, and our interactions with the environment (Capra, 1975; Wilber, 2000). By embracing the idea of interconnectedness, we can begin to see the world not as a collection of separate entities, but as a dynamic, interwoven tapestry of life and existence (Einstein, 1916; Zeilinger, 2010).

The Limitations of Human Understanding

Another key theme that has emerged is the recognition of the limitations of human understanding. Despite the remarkable advances in science and technology, there are still many aspects of the universe that remain mysterious and unexplained (Hawking, 1988). The study of black holes, quantum mechanics, and consciousness reveals the profound complexity of reality, challenging our assumptions and pushing the boundaries of what we know (Hawking, 1988; Greene, 1999).

The limitations of human understanding are not a sign of weakness, but a reminder of the vastness and mystery of the universe. They invite us to approach the study of reality with humility, acknowledging that there are likely many dimensions of existence that lie beyond our current comprehension (Hawking, 1988). This recognition encourages us to remain open to new ideas, to question our assumptions, and to be willing to revise our beliefs in light of new evidence and insights (Greene, 1999; Hawking, 1988).

At the same time, the limitations of understanding also highlight the importance of curiosity and the pursuit of knowledge. While we may never fully grasp the entirety of the universe, the process of exploration and discovery is itself a valuable and meaningful endeavor. It is through this process that we grow, learn, and evolve, both as individuals and as a species (Hawking, 1988).

In the context of spirituality, the recognition of our limitations can also be seen as an invitation to cultivate a sense of wonder and reverence for the mysteries of existence. By embracing the unknown, we can develop a deeper appreciation for the beauty and complexity of the world, and find meaning in the journey of seeking, rather than in the certainty of answers (Wilber, 2000).

The Role of Consciousness in Shaping Reality

Throughout this narrative, we have explored the idea that consciousness plays a central role in shaping our perception of reality. From the observer effect in quantum mechanics to the experiences of altered states of consciousness, there is growing evidence that the mind is not merely a passive observer of the world, but an active participant in creating the reality we experience (Bohr, 1935; Strassman, 2001).

This idea has profound implications for how we understand ourselves and our relationship with the universe. If consciousness is indeed fundamental to reality, it suggests that our thoughts, intentions, and perceptions have a direct influence on the world around us (Penrose & Hameroff, 1996). This perspective aligns with many spiritual teachings that emphasize the power of the mind, the importance of mindfulness, and the idea that we are co-creators of our own reality (Wilber, 2000).

The role of consciousness in shaping reality also invites us to consider the ethical and moral dimensions of our actions. If our thoughts and intentions have the power to influence the world, it becomes even more important to cultivate positive, compassionate, and mindful states of being (Strassman, 2001). By doing so, we can contribute to the creation of a reality that reflects our highest values and aspirations, both individually and collectively (Wilber, 2000).

The Integration of Science and Spirituality

One of the most significant insights to emerge from this exploration is the potential for integrating science and spirituality into a unified understanding of reality. While these two approaches have often been seen as opposing forces, this narrative has shown that they can complement and enrich one another, offering a more holistic perspective on the mysteries of existence (Capra, 1975; Wilber, 2000).

Science provides the tools for exploring the physical world, uncovering the laws and principles that govern the behavior of matter, energy, and space-time. It allows us to make sense of the observable universe, develop technologies that improve our lives, and push the boundaries of what is possible (Hawking, 1988).

Spirituality, on the other hand, offers a framework for exploring the inner dimensions of existence—the realms of consciousness, meaning, and connection that lie beyond the reach of empirical observation. It provides insights into the nature of the soul, the purpose of life, and the mysteries of existence that science alone cannot fully address (Wilber, 2000).

By integrating these two approaches, we can develop a more comprehensive understanding of reality—one that honors both the material and the metaphysical, the known and the unknown (Capra, 1975). This integration encourages us to embrace both reason and intuition, to seek knowledge and wisdom, and to recognize the value of both scientific inquiry and spiritual practice (Wilber, 2000).

The Ongoing Journey of Exploration

As we conclude this narrative, it is important to recognize that the journey of exploration is ongoing. The questions we have asked, the ideas we have considered, and the experiences we have reflected upon are all part of a larger, evolving process of discovery. The search for understanding, meaning, and connection is a lifelong journey—one that invites us to remain curious, open-minded, and engaged with the world around us (Capra, 1975).

In this journey, it is essential to remain flexible and adaptable, recognizing that our understanding of reality will continue to evolve as we encounter new experiences, insights, and challenges. It is also important to cultivate a sense of balance, integrating the lessons of science and spirituality, skepticism and faith, knowledge and mystery (Wilber, 2000).

Ultimately, the journey of exploration is not just about finding answers; it is about embracing the complexity and beauty of existence, and finding meaning in the process of seeking. It is about recognizing our place within the larger tapestry of the universe, and contributing to the ongoing story of life, consciousness, and reality (Capra, 1975).

Conclusion of Chapter 11 and Final Thoughts

Chapter 11 has synthesized the key themes explored throughout this narrative, reflecting on the interconnectedness of all things, the limitations of human understanding, the role of consciousness in shaping reality, and the integration of science and spirituality. These reflections offer a holistic perspective on the nature of reality and invite us to continue the journey of exploration with curiosity, humility, and wonder (Capra, 1975; Wilber, 2000).

As we move forward, let us carry with us the insights gained from this exploration, and use them to guide our search for meaning, purpose, and connection in the vast and mysterious universe we inhabit. Let us remain open to new possibilities, willing to question our assumptions, and committed to the pursuit of knowledge and wisdom in all its forms (Wilber, 2000).

This narrative is not an end, but a beginning—a starting point for further inquiry, reflection, and discovery. The mysteries of existence are deep and profound, and the journey to understand them is one that will continue to inspire and challenge us for generations to come (Capra, 1975).

Conclusion:

Embracing the Mystery and Continuing the Journey

As we reach the end of this narrative, it is clear that the journey of exploration we've embarked upon is far from over. Throughout this work, we have delved into some of the most profound and enigmatic questions that have captivated human thought for centuries: the nature of reality, the mysteries of consciousness, the interplay between science and spirituality, and the ultimate questions of life, death, and what lies beyond.

In our exploration, we have seen that the universe is a vast, interconnected web of relationships that transcends the limitations of space, time, and even our understanding. Concepts like quantum mechanics, black holes, and higher dimensions challenge our traditional views of reality, revealing a cosmos that is far stranger and more complex than we can fully comprehend. At the same time, the study of consciousness and its potential to shape reality suggests that our minds are not just passive observers of the world, but active participants in the creation of the reality we experience.

Science and spirituality, once seen as opposing forces, have emerged as complementary approaches to understanding these mysteries. Science provides the tools for exploring the physical world and uncovering the principles that govern the behavior of matter and energy. Spirituality, on the other hand, offers a framework for exploring the inner dimensions of existence—the realms of consciousness, meaning, and connection that lie beyond the reach of empirical observation. Together, these approaches offer a more holistic perspective on the nature of reality, one that honors both the material and the metaphysical, the known and the unknown.

Throughout this journey, we have also reflected on the personal experiences and insights that have shaped our understanding of these topics. Whether through the lens of mental illness, altered states of consciousness, or deep spiritual reflection, these experiences have offered valuable perspectives on the mysteries of existence. They remind us that the search for meaning and understanding is not just an intellectual pursuit, but a deeply personal and transformative journey—one that invites us to explore the boundaries of our own consciousness and to seek out the deeper connections that unite us with the universe.

As we move forward, it is important to carry with us the lessons we have learned from this exploration. The recognition of our interconnectedness with all things challenges us to live in greater harmony with the world around us, to act with compassion and empathy, and to cultivate a sense of unity and cooperation in our relationships, our communities, and our interactions with the environment. The acknowledgment of the limitations of human understanding encourages us to approach the mysteries of existence with humility and wonder, to remain open to new ideas, and to continue our pursuit of knowledge and wisdom with curiosity and dedication.

Ultimately, this narrative is not an end, but a beginning—a starting point for further inquiry, reflection, and discovery. The questions we have asked, the ideas we have considered, and the experiences we have reflected upon are all part of an ongoing quest to understand the universe and our place within it. As we continue this journey, let us remain committed to exploring the mysteries of existence with both our minds and our hearts, recognizing that the search for meaning and connection is a lifelong endeavor—one that will continue to inspire and challenge us for generations to come.

Let this narrative serve as a reminder that the universe is a profound and mysterious place, filled with wonders that beckon us to explore, to question, and to seek out the deeper truths that lie at the heart of existence. As we embrace the mystery and continue our journey, may we find the courage to venture into the unknown, the wisdom to learn from our experiences, and the compassion to share what we have discovered with others. In doing so, we honor the timeless quest for understanding that unites us all, and we contribute to the ongoing story of life, consciousness, and reality in this vast and beautiful universe.

As we reach the end of this exploration, it's clear that the boundaries between science, spirituality, and the mysteries of the universe are not as rigid as they might seem. This work has been an attempt to challenge conventional perspectives, to delve into the unknown, and to consider possibilities that stretch beyond the limits of our current understanding.

Throughout this journey, I have refrained from promoting any specific religion or belief system. I believe it's not my place to tell anyone what they should center their beliefs around. My own belief system is not static; it evolves as I continue to seek understanding of the larger picture. I recognize that many forms of spirituality can offer valuable insights and benefits, and that spirituality can be personal and unique to each individual. Even those who do not identify with any religion, including those who consider themselves atheists, might find value in admitting that there could be aspects of existence beyond our present comprehension.

In closing, I hope that this work encourages you to think deeply, question your assumptions, and explore the mysteries that lie at the intersection of science and spirituality. The universe is vast and complex, and our understanding of it is always growing. By embracing curiosity and open-mindedness, we can continue this journey of discovery together.

APPENDICES

Appendix A: Glossary of Key Terms

The Universe:
- **Description:** The totality of all space, time, matter, and energy. It includes everything that exists, from the smallest subatomic particles to the largest galaxies, as well as the physical laws and constants that govern them. The universe is often studied in cosmology, which seeks to understand its origin, structure, and eventual fate.

Field of Science:
- **Description:** The systematic study of the natural world through observation, experimentation, and evidence-based reasoning. Science seeks to understand the laws and principles that govern the universe, from the smallest particles to the largest structures.

Field of Spirituality:
- **Description:** The study and practice of beliefs, values, and experiences related to the human spirit, the soul, and the search for meaning and connection with something greater than oneself. Spirituality often involves exploring questions about existence, purpose, and the nature of the divine.

Quantum Mechanics:
- **Description:** A branch of physics that deals with the behavior of particles at the smallest scales, where the laws of classical physics do not apply. It explores phenomena like wave-particle duality, quantum entanglement, and the uncertainty principle.

Black Hole:
- **Description:** A region of space where gravity is so strong that nothing, not even light, can escape from it. Black holes are often formed when a massive star collapses at the end of its life cycle.

Singularity:
- **Description:** A point at the center of a black hole where the laws of physics break down and densities become infinite. It represents a point where current physical theories cannot describe the conditions.

Event Horizon:
- **Description:** The boundary around a black hole beyond which no information or matter can escape. It is the point of no return for anything falling into a black hole.

Consciousness:
- **Description:** The state of being aware of and able to think, perceive, and experience. Consciousness remains one of the most profound mysteries in both science and philosophy.

Altered State of Consciousness:
- **Description:** A condition in which the normal functioning of consciousness is temporarily altered, leading to a different perception of reality. Altered states of consciousness can be induced by various means, including meditation, psychedelics like DMT, sleep deprivation, or intense emotional experiences. These states often involve changes in perception, thought patterns, and a sense of self.

Bipolar Disorder:

- **Description:** A mental health condition characterized by extreme mood swings, including periods of intense highs (manic episodes) and lows (depressive episodes). These mood swings can affect energy levels, behavior, and the ability to carry out daily tasks.

Manic Episode:
- **Description:** A period of abnormally elevated or irritable mood, heightened energy, and hyperactivity, often associated with bipolar disorder. During a manic episode, individuals may experience grandiose thoughts, decreased need for sleep, and impulsive behavior.

DMT (Dimethyltryptamine):
- **Description:** A naturally occurring psychedelic compound found in many plants and animals. DMT is known for producing intense and often transformative experiences, including vivid visions and a sense of encountering otherworldly beings. It is sometimes called the "spirit molecule" due to its powerful effects and its role in many cultural and spiritual practices.

NDE (Near-Death Experience):
- **Description:** A profound personal experience associated with death or impending death, often involving sensations such as leaving the body, traveling through a tunnel, encountering deceased loved ones, and a sense of peace. NDEs are frequently reported by individuals who have been close to death and subsequently revived.

Retrocausality:

- **Description:** A concept in quantum mechanics where an effect precedes its cause, challenging traditional notions of time and causality. In the context of the Double-Slit Experiment, it refers to how a decision made after particles pass through slits can retroactively determine their behavior, suggesting that future events can influence the past.

Observer Effect:

- **Description:** The principle in quantum mechanics that the act of observation can alter the behavior of particles. This phenomenon is famously demonstrated in the Double-Slit Experiment, where particles act as waves when unobserved but behave like particles when observed.

Quantum Entanglement:

- **Description:** A phenomenon in quantum mechanics where two or more particles become linked, such that the state of one particle instantaneously influences the state of the other, regardless of the distance separating them. This challenges classical ideas about locality and information transfer.

Wave-Particle Duality:

- **Description:** A fundamental concept in quantum mechanics that posits particles, such as electrons and photons, can exhibit both wave-like and particle-like properties depending on the experimental setup. This duality is a key feature of the Double-Slit Experiment.

Transactional Interpretation:

- **Description:** An interpretation of quantum mechanics that posits particles can send and receive signals across time, engaging in "handshakes" between the past and future to establish the outcomes of quantum events. This interpretation provides a framework for understanding retrocausality.

Binary Code:

- **Description:** A system of representing text or computer processor instructions using the binary number system's two binary digits, 0 and 1. In the context of "101," binary code represents the number 5, which can symbolize balance, adaptability, and dynamic change.

Angel Number 101:

- **Description:** In numerology and spiritual traditions, the number "101" is considered an "Angel Number," symbolizing new beginnings, spiritual awakening, and the cyclical nature of life. It is believed to represent guidance from higher spiritual forces, encouraging individuals to trust in their path and embrace new opportunities.

Appendix B: Further Reading and Resources

Books:

- *"The Universe in a Nutshell"* by Stephen Hawking – A comprehensive guide to modern theoretical physics, including discussions on black holes and quantum mechanics.
- *"The Fabric of Reality"* by David Deutsch – Explores how quantum theory, parallel universes, and other ideas are interwoven to form the fabric of reality.
- *"The Quantum Enigma: Physics Encounters Consciousness"* by Bruce Rosenblum and Fred Kuttner – A book that delves into the connection between quantum mechanics and consciousness.
- *"Death by Black Hole: And Other Cosmic Quandaries"* by Neil deGrasse Tyson – A collection of essays that covers various astrophysical phenomena, including black holes, with clarity and wit, making complex scientific ideas accessible to a broad audience.

Articles:

- *"The Holographic Principle and Black Holes"* – An overview of the holographic principle and how it relates to black holes and the preservation of information.

- *"Quantum Mechanics and the Observer Effect"* – An article discussing the role of the observer in quantum mechanics and its implications for reality.

Documentaries:

- *"Cosmos: A Spacetime Odyssey"* – A science documentary series hosted by Neil deGrasse Tyson that explores the vastness of the universe, the laws of nature, and our place in the cosmos.
- *"Through the Wormhole"* – A documentary series hosted by Morgan Freeman that examines the deepest mysteries of existence, including black holes, consciousness, and the nature of reality.
- *"The Universe" (History Channel Series)* – A documentary series that covers a wide range of topics related to cosmology, including black holes and the structure of the universe.
- *"How the Universe Works"* – This series provides a comprehensive look at the universe, explaining the science behind the formation of galaxies, the life cycle of stars, black holes, and other cosmic phenomena, with insights from leading scientists and astrophysicists.

Appendix C: Overview of Scientific Theories

String Theory:

- **Description:** String theory posits that the fundamental particles of the universe are not point-like dots but rather tiny, vibrating strings. The different vibrations of these strings give rise to different particles, potentially unifying all the fundamental forces and particles in the universe.

- **Relevance:** String theory is one of the leading candidates for a unified theory because it naturally includes gravity along with the other fundamental forces. It also introduces the idea of extra dimensions, which might explain phenomena that are currently beyond our understanding.

M-Theory:

- **Description:** M-Theory is an extension of string theory that proposes that strings are actually one-dimensional slices of a two-dimensional membrane vibrating in an 11-dimensional space. M-Theory attempts to unify the five different string theories into a single framework.

- **Relevance:** M-Theory is considered by some physicists to be the ultimate theory of everything, potentially explaining all the forces of nature, including gravity, within a single coherent framework.

Grand Unified Theory (GUT):

- **Description:** Grand Unified Theories aim to unify the three fundamental forces of the Standard Model (electromagnetic, weak, and strong nuclear forces) into a single force. GUTs are a step toward a more comprehensive Theory of Everything but do not include gravity.
- **Relevance:** GUTs are significant in the journey towards a complete unified theory, as they represent an effort to bridge the gap between the Standard Model and the inclusion of gravity in a single framework.

Theory of Everything (TOE):

- **Description:** The Theory of Everything is a hypothetical framework that seeks to fully explain and link together all physical aspects of the universe. It would encompass all known fundamental interactions in a single theory, unifying quantum mechanics, general relativity, and potentially other forces.
- **Relevance:** A TOE would be the ultimate goal of theoretical physics, providing a single equation or set of equations that describe all physical phenomena, from the behavior of subatomic particles to the expansion of the universe.

The Holographic Principle:

- **Description:** The Holographic Principle suggests that all the information contained within a volume of space can be represented as a theory on the boundary of that space. In the context of black holes, this principle implies that the information about everything that falls into a black hole might be encoded on its event horizon rather than being lost inside.

- **Relevance:** The Holographic Principle offers a potential resolution to the information paradox by suggesting that information is not destroyed but rather stored on the event horizon of the black hole.

Black Hole Complementarity:

- **Description:** Black Hole Complementarity posits that information about matter that falls into a black hole is preserved both on the event horizon and inside the black hole, depending on the observer's point of view. This theory suggests that the paradox arises because different observers perceive the information differently, but no information is actually lost.

- **Relevance:** Black Hole Complementarity offers a way to reconcile the information paradox without violating quantum mechanics, by suggesting that both perspectives are valid and complementary.

Firewalls (AMPS Paradox):

- **Description:** The firewall hypothesis suggests that an energetic "firewall" forms at the event horizon of a black hole, which would incinerate anything falling in. This idea challenges the concept of black hole complementarity and suggests a new paradox.

- **Relevance:** The firewall paradox challenges existing resolutions to the information paradox by proposing that preserving information might require the breakdown of the event horizon as a smooth boundary.

Soft Hair on Black Holes:
- **Description:** Proposed by Stephen Hawking and others, this theory suggests that black holes might have "soft hair" — low-energy quantum excitations on the event horizon that store information, preventing its loss.
- **Relevance:** The concept of "soft hair" provides a mechanism by which information might be stored and preserved on the black hole's event horizon, offering a potential solution to the information paradox.

Quantum Entanglement and the ER=EPR Conjecture:
- **Description:** The ER=EPR conjecture suggests that quantum entanglement might be connected to Einstein-Rosen bridges (wormholes), implying that entangled particles could be connected by non-traversable wormholes, which could help preserve information in black holes.
- **Relevance:** This conjecture offers a novel approach to the information paradox, suggesting that entanglement could help preserve information across space-time, even if it appears lost.

The Multiverse Hypotesis:
- **Description:** The idea that our universe is just one of many, potentially infinite, universes that exist in parallel with one another.

- **Relevance:** The Multiverse Hypothesis challenges the notion of a singular universe, opening up possibilities for different physical laws and realities in other universes.

Hawking Radiation:

- **Description:** Theoretical radiation predicted by Stephen Hawking that is emitted by black holes, suggesting that they can slowly lose mass and eventually evaporate.
- **Relevance:** Hawking Radiation provides insights into the behavior of black holes, connecting quantum mechanics with general relativity and offering clues towards a unified theory.

Simulation Hypothesis:

- **Description:** The Simulation Hypothesis proposes that reality, as we perceive it, might be an artificial simulation, such as a computer-generated environment. This theory suggests that the universe could be "programmed" to behave differently based on observation, similar to how a video game only renders what the player can see.
- **Relevance:** The Simulation Hypothesis challenges our understanding of reality by suggesting that our universe might be an artificial construct. It opens up philosophical and scientific inquiries into the nature of existence and consciousness.

Collective Consciousness:

- **Description:** Collective Consciousness is the theory that individual consciousnesses are interconnected, potentially influencing one another and contributing to a shared experience or understanding of reality.

- **Relevance:** Collective Consciousness offers a framework for understanding how individual experiences and observations might contribute to a unified, collective reality. It has implications for theories in quantum mechanics, psychology, and sociology.

Anthropic Principle:

- **Description:** The Anthropic Principle is a philosophical consideration that the universe's fundamental constants are fine-tuned to allow for the existence of life. It suggests that the conditions observed in the universe must be compatible with the conscious beings observing it, potentially hinting at a deeper connection between consciousness and the cosmos.
- **Relevance:** The Anthropic Principle plays a significant role in discussions about the nature of the universe and the possibility of multiple universes or higher-order design.

Appendix D: Explanation of Key Experiments and Thought Experiments

The Double-Slit Experiment:

- **Explanation:** An experiment in quantum mechanics that demonstrates the wave-particle duality of light and electrons, and the role of the observer in determining the outcome. It shows how particles like electrons can behave both as particles and as waves, depending on whether they are observed, challenging the classical notion of deterministic behavior.

The LIGO Experiment:

- **Explanation:** The Laser Interferometer Gravitational-Wave Observatory (LIGO) detects gravitational waves—ripples in spacetime caused by violent astronomical events like black hole collisions. This experiment confirmed Einstein's predictions and provided a new way to observe the universe.

The Flatland Thought Experiment:

- **Explanation:** Flatland: A Romance of Many Dimensions is a novella by Edwin A. Abbott, published in 1884, that explores the concept of higher dimensions. The story is set in a two-dimensional world inhabited by geometric shapes. The protagonist, a square, encounters a three-dimensional sphere, but struggles to comprehend the idea of a third dimension. This thought experiment illustrates how beings limited to lower dimensions might struggle to understand higher dimensions. It serves as a metaphor for our potential limitations in perceiving the full reality of the universe, suggesting that there could be aspects of existence beyond our current understanding.

The EPR Paradox and Bell's Theorem:

- **Explanation:** The Einstein-Podolsky-Rosen (EPR) Paradox, proposed in 1935, questioned the completeness of quantum mechanics by suggesting that the theory implies "spooky action at a distance," where two particles can instantaneously affect each other's state, no matter the distance between them. This paradox led to Bell's Theorem in 1964, which provided a way to test the paradox experimentally. Bell's experiments have shown that quantum entanglement does indeed occur, challenging classical notions of locality and realism, and confirming the non-local nature of quantum mechanics.

The Schrödinger's Cat Thought Experiment:

- **Explanation:** Schrödinger's Cat is a thought experiment devised by physicist Erwin Schrödinger in 1935 to illustrate the strange nature of quantum superposition. In this experiment, a cat is placed in a sealed box with a radioactive atom that has a 50% chance of decaying and releasing poison, which would kill the cat. Until the box is opened and the cat is observed, the cat is considered to be both alive and dead simultaneously. This paradox highlights the problems of interpreting quantum mechanics and questions when and how quantum superposition ends and reality collapses into one of the possible states.

The Trolley Problem:

- **Explanation**: The Trolley Problem is a philosophical thought experiment introduced by Philippa Foot in 1967, exploring moral dilemmas and ethical decision-making. In its basic form, the experiment presents a scenario where a trolley is headed towards five people tied to a track. You have the option to pull a lever, diverting the trolley onto another track where only one person is tied. The problem raises questions about utilitarianism, the morality of action versus inaction, and the value of human life.

The Black Hole Information Paradox:

- **Explanation**: Proposed by Stephen Hawking in the 1970s, the Black Hole Information Paradox arises from the conflict between quantum mechanics and general relativity. According to quantum mechanics, information must be conserved, but Hawking's theory of black hole radiation suggests that information could be lost forever in black holes, violating this principle. This paradox has led to extensive research and debate, with various proposed solutions, including the idea of holographic principles or the notion that information is somehow preserved in black hole remnants or radiation.

The Many-Worlds Interpretation:

- **Explanation**: The Many-Worlds Interpretation, proposed by Hugh Everett in 1957, is a thought experiment and interpretation of quantum mechanics that suggests every quantum event results in a branching of the universe into multiple, parallel worlds, each representing different outcomes. This interpretation challenges the traditional view of a single, objective reality and proposes that all possible outcomes of a quantum measurement actually occur, each in its own separate universe.

The Michelson-Morley Experiment:

- **Explanation**: Conducted in 1887 by Albert A. Michelson and Edward W. Morley, this experiment aimed to detect the presence of aether, a medium through which light was once thought to travel. The experiment found no evidence of aether, leading to the eventual abandonment of the aether theory and contributing to the development of Einstein's theory of special relativity, which revolutionized our understanding of space and time.

The Chinese Room Thought Experiment:

- **Explanation**: Proposed by philosopher John Searle in 1980, the Chinese Room Thought Experiment challenges the notion of strong artificial intelligence (AI) by questioning whether a computer running a program can possess a "mind" or "understanding." Searle imagines a person inside a room following instructions to manipulate Chinese symbols, producing responses indistinguishable from a native speaker.

Despite this, the person does not understand Chinese. The experiment argues that mere symbol manipulation (as done by computers) is not equivalent to true understanding or consciousness.

Appendix E: References

- Armstrong, K. (2006). The Great Transformation: The Beginning of Our Religious Traditions. Knopf.
- Armstrong, K. (2010). Twelve Steps to a Compassionate Life. Knopf.
- Barad, K. (2007). Meeting the Universe Halfway: Quantum Physics and the Entanglement of Matter and Meaning. Duke University Press.
- Barbour, I. G. (1997). Religion and Science: Historical and Contemporary Issues. HarperCollins.
- Barrow, J. D., & Tipler, F. J. (1986). The Anthropic Cosmological Principle. Oxford University Press.
- Bergson, H. (1934). The Creative Mind: An Introduction to Metaphysics. Philosophical Library.
- Bhagavad Gita. (n.d.). The Bhagavad Gita. (Any standard edition can be used for citation).
- Bohr, N. (1935). Can quantum-mechanical description of physical reality be considered complete? Physical Review, 48(8), 696-702.
- Bostrom, N. (2003). Are You Living in a Computer Simulation? Philosophical Quarterly, 53(211), 243-255.
- Bousso, R. (2002). The holographic principle. Reviews of Modern Physics, 74(3), 825-874.
- Capra, F. (1975). The Tao of Physics: An Exploration of the Parallels Between Modern Physics and Eastern Mysticism. Shambhala Publications.

- Chödrön, P. (1997). When Things Fall Apart: Heart Advice for Difficult Times. Shambhala Publications.
- Corrigan, P. W., & Watson, A. C. (2002). Understanding the impact of stigma on people with mental illness. World Psychiatry, 1(1), 16-20.
- Dawkins, R. (2006). The God Delusion. Bantam Press.
- Deloria, V. (1994). God is Red: A Native View of Religion. Fulcrum Publishing.
- Deutsch, D. (1997). The Fabric of Reality: The Science of Parallel Universes—and Its Implications. Penguin Books.
- Einstein, A. (1916). The foundation of the general theory of relativity. Annalen der Physik, 49(7), 769-822.
- Einstein, A. (1926). Letter to Max Born. In M. Born (Ed.), The Born-Einstein Letters: Friendship, Politics and Physics in Uncertain Times (translated by I. Born). Macmillan.
- Eliade, M. (1987). The Sacred and the Profane: The Nature of Religion. Harcourt.
- Ellis, J. (2004). The Search for Unity: The Quest for Global Peace and Universal Harmony. New World Library.
- Feyerabend, P. (1975). Against Method. Verso.
- Feynman, R. P. (1985). QED: The Strange Theory of Light and Matter. Princeton University Press.
- Frankl, V. E. (1946). Man's Search for Meaning. Beacon Press.
- Fuller, B. (1969). Operating Manual for Spaceship Earth. Southern Illinois University Press.

- Gallimore, A. R. (2015). Restructuring consciousness: The psychedelic state in light of integrated information theory. Frontiers in Human Neuroscience, 9, 346.
- Gardner, H. (1983). Frames of Mind: The Theory of Multiple Intelligences. Basic Books.
- Goleman, D. (1995). Emotional Intelligence: Why It Can Matter More Than IQ. Bantam Books.
- Goodwin, F. K., & Jamison, K. R. (2007). Manic-Depressive Illness: Bipolar Disorders and Recurrent Depression (2nd ed.). Oxford University Press.
- Greene, B. (1999). The Elegant Universe: Superstrings, Hidden Dimensions, and the Quest for the Ultimate Theory. W.W. Norton & Company.
- Greyson, B. (1983). The near-death experience scale: Construction, reliability, and validity. Journal of Nervous and Mental Disease, 171(6), 369-375.
- Hanh, T. N. (1998). The Heart of the Buddha's Teaching. Broadway Books.
- Hawking, S. W. (1974). Black hole explosions? Nature, 248(5443), 30-31.
- Hawking, S. W. (1988). A Brief History of Time: From the Big Bang to Black Holes. Bantam Books.
- Heidegger, M. (1927). Being and Time. Harper & Row.
- James, W. (1897). The Will to Believe and Other Essays in Popular Philosophy. Longmans, Green, and Co.
- Jamison, K. R. (1993). Touched with Fire: Manic-Depressive Illness and the Artistic Temperament. Free Press.

- Jung, C. G. (1963). Memories, Dreams, Reflections. Pantheon Books.
- Kant, I. (1781). Critique of Pure Reason. Cambridge University Press.
- Kerr, R. P. (1963). Gravitational field of a spinning mass as an example of algebraically special metrics. Physical Review Letters, 11(5), 237-238.
- King, M. L. (1963). Strength to Love. Harper & Row.
- Knuth, D. E. (1997). The Art of Computer Programming, Volume 1: Fundamental Algorithms (3rd ed.). Addison-Wesley.
- Krishnamurti, J. (1969). Freedom from the Known. Harper & Row.
- Kuhn, T. S. (1962). The Structure of Scientific Revolutions. University of Chicago Press.
- Laszlo, E. (2004). Science and the Akashic Field: An Integral Theory of Everything. Inner Traditions.
- Macy, J. (1991). World as Lover, World as Self: Courage for Global Justice and Ecological Renewal. Parallax Press.
- Mayr, E. (1982). The Growth of Biological Thought: Diversity, Evolution, and Inheritance. Harvard University Press.
- Merikangas, K. R., Jin, R., He, J. P., Kessler, R. C., Lee, S., Sampson, N. A., ... & Zarkov, I. (2011). Prevalence and correlates of bipolar spectrum disorder in the world mental health survey initiative. Archives of General Psychiatry, 68(3), 241-251.

- Meyer, M. (2016). Exploring the Mind: The Philosophy and Science of Consciousness. Routledge.
- Moody, R. A. (1975). Life After Life: The Investigation of a Phenomenon—Survival of Bodily Death. HarperCollins.
- OpenAI. (2024). ChatGPT (August 16 version). OpenAI. https://www.openai.com/chatgpt
- Ostrom, E. (1990). Governing the Commons: The Evolution of Institutions for Collective Action. Cambridge University Press.
- Page, D. N. (1993). Information in black hole radiation. Physical Review Letters, 71(23), 3743-3746.
- Pais, A. (1982). Subtle is the Lord: The Science and the Life of Albert Einstein. Oxford University Press.
- Penrose, R. (1989). The Emperor's New Mind: Concerning Computers, Minds, and the Laws of Physics. Oxford University Press.
- Penrose, R., & Hameroff, S. (1996). Shadows of the Mind: A Search for the Missing Science of Consciousness. Oxford University Press.
- Popper, K. (1934). The Logic of Scientific Discovery. Routledge.
- Preskill, J. (1992). Do black holes destroy information? International Journal of Modern Physics D, 1(3), 145-166.
- Randall, L., & Sundrum, R. (1999). An alternative to compactification. Physical Review Letters, 83(23), 4690-4693.

- Robinson, K. (2015). Creative Schools: The Grassroots Revolution That's Transforming Education. Viking.
- Sartre, J. P. (1943). Being and Nothingness: An Essay on Phenomenological Ontology. Washington Square Press.
- Shanon, B. (2002). *The Antipodes of the Mind: Charting the Phenomenology of the Ayahuasca Experience.* Oxford University Press.
- Sheldrake, R. (1981). A New Science of Life: The Hypothesis of Formative Causation. Blond & Briggs.
- Sheldrake, R. (2012). The Science Delusion: Freeing the Spirit of Enquiry. Coronet.
- Singer, P. (1981). The Expanding Circle: Ethics and Sociobiology. Farrar, Straus and Giroux.
- Smith, H. (2009). Forgotten Truth: The Common Vision of the World's Religions. HarperOne.
- Strassman, R. (2001). DMT: The Spirit Molecule: A Doctor's Revolutionary Research into the Biology of Near-Death and Mystical Experiences. Park Street Press.
- Susskind, L. (2008). The Black Hole War: My Battle with Stephen Hawking to Make the World Safe for Quantum Mechanics. Little, Brown and Company.
- Thorne, K. S. (1994). Black Holes and Time Warps: Einstein's Outrageous Legacy. W. W. Norton & Company.
- Tutu, D. (1999). No Future Without Forgiveness. Doubleday.
- Virtue, D. (2005). *Angel Numbers 101: The Meaning of 111, 123, 444, and Other Number Sequences.* Hay House, Inc.

- von Neumann, J. (1955). Mathematical Foundations of Quantum Mechanics. Princeton University Press.
- West, D. (2011). Numerology: Your Personal Guide for Life. Hay House, Inc.
- Whitehead, A. N. (1929). Process and Reality: An Essay in Cosmology. Free Press.
- Wilber, K. (1998). The Marriage of Sense and Soul: Integrating Science and Religion. Random House.
- Wilber, K. (2000). A Theory of Everything: An Integral Vision for Business, Politics, Science and Spirituality. Shambhala Publications.
- Wheeler, J. A. (1978). The 'past' and the 'delayed-choice' double-slit experiment. In A.R. Marlow (Ed.), Mathematical Foundations of Quantum Theory. Academic Press.
- Wilson, E. O. (2002). The Creation: An Appeal to Save Life on Earth. W.W. Norton & Company.
- Zeilinger, A. (2010). Dance of the Photons: From Einstein to Quantum Teleportation. Farrar, Straus and Giroux.

Acknowledgments

I would like to express my deepest gratitude to my family, who have been a constant source of encouragement and understanding during struggles with mental disablitity. They have been especially helpful in supporting me to maintain a normal life despite the challenges of my mental illness. I would particularly like to thank my wife, who has provided me with a stability I never thought possible given my condition. I also would thank my daughter who has taught me so much about the miracle of life and has inspired me to be better with every passing day.

I am also grateful to the medical professionals who have shown great understanding and have helped me navigate the complexities of living with bipolar disorder. Their care and guidance have been invaluable.

This work has been profoundly influenced by the writings and research of those who explore the frontiers of science and consciousness. Their dedication to uncovering the mysteries of the universe has been a continual source of inspiration.

To all those unnamed individuals who have contributed to my journey, whether through direct support or by inspiring my thoughts and ideas, I offer my sincerest thanks.

About the Author

Pearz Reagan is a writer and researcher with a diverse academic and professional background. With a BS in Biological Engineering and an MBA, Pearz is currently pursuing advanced degrees in Financial Engineering and Artificial Intelligence. This unique blend of disciplines informs Pearz's interdisciplinary exploration of science, consciousness, and the hidden aspects of reality.

Pearz's career has spanned various fields, including transgenetic crop research, food manufacturing engineering, and data analysis in nonprofit early childhood development programs. Growing up on a family farm in the Midwest, Pearz's humble beginnings have deeply influenced a lifelong curiosity about the natural world and the forces that shape it.

For over 15 years, Pearz has navigated the complexities of living with bipolar disorder, an experience that has profoundly shaped both personal insights and professional endeavors. As an advocate for the rights of those with mental disabilities, Pearz draws from personal experiences with mental illness to raise awareness and promote understanding.

Through writing, Pearz seeks to challenge conventional perspectives and explore the boundaries of human understanding, inviting readers to journey into the unknown and consider the profound interconnectedness of all things.

Contact Information

I would love to hear your feedback on the topics discussed in this book. Whether you have questions, thoughts, or would simply like to share your perspective, I welcome correspondence from readers. Additionally, I am open to speaking engagements where we can delve deeper into these fascinating subjects.

- **Email:** pearz.reagan@engineer.com
- **Website:** www.pearzreagan.life

For those interested in exploring these topics further, I invite you to engage with a specially designed GPT that can help deepen your understanding and stimulate thought-provoking discussions:

- **Explore GPT:**

    https://chatgpt.com/g/g-Pa3Zsfd46-conceptualizing-the-unseen-universe